SUGGESTIONS FOR TEACHING ELECTRICAL AND BASIC CONTROLS

A Hobar Publications Workbook

First Globe Pequot edition 2019

Published by Hobar Publications
An imprint of Globe Pequot Press
Wholly Owned by: The Rowman & Littlefield Publishing Group, Inc.
4501 Forbes Boulevard, Suite 200
Lanham, Maryland 20706

Distributed by National Book Network
1-800-462-6420

ACKNOWLEDGMENTS

Major credit in securing permission from the Texas Farm Electrification Committee to edit and publish this set of electrical controls lesson plans for vocational agriculture teachers should go to R. H. Browder, EEI Youth Committee Chairman, Texas Electric Service Company. The assistance of the following Youth Committee members is also appreciated:

C. A. Bradford, Public Service Company of Oklahoma; P. A. Cochran, Delmarva Power & Light Company of Delaware; J. L. Copeman, Monongahela Power Company; E. R. Doughty, Central Maine Power Company; L. R. Hansen, Pacific Power & Light Company; Nathan Haw, Northern States Power Company; J. L. Hays, West Texas Utilities Company; E. R. Heacock, Central Illinois Public Service Company; A. E. Kettelhut, The Cincinnati Gas & Electric Company.

Special recognition for their contributions to this publication should be given to the following:

Mr. John M. Spence, Senior Agricultural Engineer, Alabama Power Company, Birmingham, Alabama

Mr. Wayne D. Jones, Rural Engineer, Illinois Power Company, Decatur, Illinois

Professor J. W. Matthews, Vocational-Agriculture Service, 434 Mumford Hall, University of Illinois, Urbana, Illinois

Mr. Robert C. Jaska, Coordinator, Adult Specialist Program, Texas Education Agency, College Station, Texas

Mr. W. E. McCune, Director, Texas Farm Electrification Committee

Members of the Youth Committee of the Texas Farm Electrification Committee

SUGGESTIONS FOR TEACHING ELECTRICAL AND BASIC CONTROLS USED IN AGRICULTURAL PRODUCTION

published by

EDISON ELECTRIC INSTITUTE
750 Third Avenue
New York, New York 10017

FOREWORD

The lesson plans in this booklet have been prepared to assist in carrying out vocational and technical instruction programs that will meet the needs of adult, FFA, 4-H, and other youth groups in teaching the fundamentals of electrical controls and control circuits.

The material was compiled by staff members of the Agricultural Engineering and Agricultural Education Departments of Texas A & M University, with the cooperation of vocational agriculture teachers and the investor owned power companies in Texas.

Electrical controls and basic control circuits, when applied to agricultural production, can provide both comfort and convenience as well as save time and labor. For this reason these lesson plans have been designed to assist teachers and others in carrying out efficient instruction in this area. It is intended that this publication be used only as a guide.

The Farm Group of Edison Electric Institute recommends use of this teaching guide and wishes to express appreciation to the Texas Farm Electrification Committee for its courtesy in permitting Edison Electric Institute to publish and distribute this material.

TABLE OF CONTENTS

Page

B. Study of relay

    1. Ampere rating
    2. Voltage rating
    3. Motor rating
    4. Number of poles - normally open or normally closed

C. Study of relay circuits

    1. Circuit components
    2. Load circuit
    3. Control circuit

D. Exercises and skills

    1. Connect control circuit components
    2. Energize circuits and operate
    3. Show application of controls and circuits

E. Time delay relays

    1. Purpose
    2. Construction and operation
    3. Application in control circuits

Lesson Plan IV - 1 hour

Motor Control Devices

A. Purpose
B. Manual controller
C. Magnetic motor controller

    1. Components
    2. Rating
    3. Overload protection

D. Study of magnetic starter and schematic diagrams
E. Push button motor control stations

    1. Purpose
    2. Rating
    3. Study of push button stations and schematic wiring diagrams

F. Exercises and skills

Lesson Plan V - 2 hours

Automatic Sensing Control Devices

A. Purpose
B. Temperature controllers - operation and use in control circuits
C. Humidity controllers - operation and use in control circuits
D. Timing devices - application of timing devices in control circuits
E. Photoelectric cells - operation and use in control circuits

Suggestions for Teaching

Electrical and Basic Control Circuits

Used in Agricultural Production

COURSE OUTLINE

Lesson Plan I - 1 hour

Introduction

A. Farm Automation
B. Purpose and function of control devices
C. Non-automatic control systems
D. Automatic control systems
E. Terms and concepts

   1. Switches - poles and throw
   2. Switches - normally open and normally closed
   3. Schematic diagrams

F. Exercises and skills

Lesson Plan II - 2 hours

Switches and Switch Control Circuits

A. Study of switches

   1. Ampere rating
   2. Voltage rating
   3. Motor rating
   4. Number poles - normally open or normally closed

B. Use of tumbler or toggle switches
C. Use of switches with built-in overload protection
D. Use of limit switches
E. Use of 3-way and 4-way switches
F. Exercises and skills

   1. Study switch control circuits
   2. Draw control circuit diagrams

Lesson Plan III - 2 hours

Relay Devices

A. Introduction
   1. Purpose
   2. Components
   3. Operation

Farm automation, as we know it today, depends upon the extensive use of time switches, thermostats, pressure switches, humidistats, and other similar devices which start and stop electrical equipment without the constant attention of the farm operator. Farm automation goes beyond "pushbutton" farming. Automatic controls make it unnecessary even to push a button.

B. The basic purpose of electrical control devices is for starting and stopping electrical equipment.

In performing their basic function, controls also provide:

1. Safety (list several - fuses, circuit breakers, etc.)

2. Convenience (list several - time clocks, photoelectric cells, etc.)

3. Desired environmental or comfort conditions - (list several - thermostat, humidistat, heaters, etc.)

4. Desired type motion (list several - switches, motor starters, relays, etc.)

C. Non-automatic control devices:

1. The simplest control devices are those that are not automatic, but require the attention of an operator for starting and stopping. The attachment plug on the end of the cord of an electric appliance is one type of simple control.

2. Another type of relatively simple control for fractional horsepower motors is the common snap-action type tumbler switch which is connected permanently in the motor circuit (Fig. 1). This type switch is hand operated, meaning that the operator must snap the switch manually. Contact points within the switch itself close to allow current to flow, or open to break the circuit.

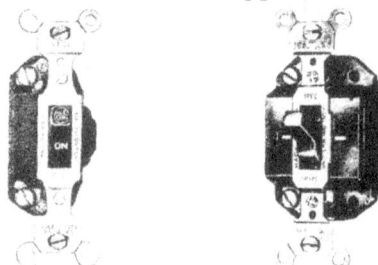

Fig. 1. Snap-action type switch. This type switch provides no overload protection.

Some snap-action switches contain a short circuit or overload protection device for protecting electric motors (Fig. 2). These switches are designed for starting and protecting small single phase motors.

Fig. 2. Snap-action switch with overload protection. The overload device is made in many ampere sizes.

10

INTRODUCTION TO ELECTRIC CONTROLS AND CONTROL CIRCUITS

This lesson plan is designed to be covered in a 1-hour period

TEACHING OBJECTIVES:

1. To help the student to understand the function and importance of controls and control circuits in the operation of electric equipment.

2. To familiarize the student with the terminology and symbols used in discussing electric control circuits.

3. To teach the student the characteristics of automatic and non-automatic control systems.

TEACHING AIDS:

1. Tools and Equipment

    Simple attachment plugs of different types - (3)
    Toggle switch - (1)
    Snap action tumbler switch - (1)
    Snap action switch with overload protection - (1)
    Magnetic relay - (1)
    Thermostat, humidistat, or time clock - (1)
    Small motor or socket and light bulb - (1)
    Connection wires

2. Visual Aids

    Blackboard
    Opaque projector
    Overhead projector and screen
    Transparencies:
        TP 1-1EC - Schematic diagram of single pole, double pole and three pole switches
        TP 2-1EC - Schematic diagram of SPST, DPST, SPDT, DPDT & 3PDT switches
        TP 3-1EC - Schematic diagram of normally open and normally closed switches

3. Sources of Information

    Manufacturers' catalogs - (Square D, Allen-Bradley, Westinghouse, etc.)
    Myers Prefiled Agricultural Catalogs

PROCEDURE:

A. One good reason for the acceptance of electrical equipment on farms is that it can be easily controlled and adapted to automatic operation.

quite likely would be impractical for a farmer to get out of bed at that time to do the same job.

A humidistat can actuate a ventilation fan at any desired humidity setting. A man is generally not capable of recognizing limited humidity changes. Neither can a man recognize changes in a pressure water system, at least in the ranges generally required, but a pressure switch can sense the change.

Some sensing elements have sufficient current capacity to operate a motor directly. For example, a pressure switch on a farmstead water system is generally connected directly to the pump motor.

For some motor operations, it is desirable to have the sensing element connected to a control device called a magnetic relay. The relay, when energized, permits current to flow to the motor. This method of control oftentimes enables a low voltage, such as 12 or 24 volts, to be used from the sensing element to the energizing coil of the relay. The external contact points of the relay then close to permit current from the power source of 120 or 240 volts to flow to the motor.

The contact points on a sensing element would burn away if connected directly to a large motor. The contacts of the sensing element operating on 12 or 24 volts will not be burned. The relay contact points would carry the full motor current and would have sufficient capacity to permit the full voltage and the full rated current.

A common example is a thermostat controlling an air conditioner. This control system requires a transformer to reduce 120 to 24 volts, a low voltage thermostat which is the sensing element, a relay, and compressor motor. When the temperature gets to a preset figure, the thermostat permits current to flow to the relay. The relay closes to complete the circuit and permits the compressor motor to run.

There are other advantages to such a control system. Small, inexpensive wire can be used to and from the thermostat since the control circuit carries only the current required to operate the relay. Also, the thermostat need not be so well insulated and, hence, is less expensive. Low voltage controls and control circuits are also desirable from a safety standpoint.

E. Terms and concepts

A better knowledge of the various switches and control devices discussed thus far will be gained by laboratory exercises. It is necessary, however, to first become familiar with some basic terms and ideas before you can fully understand the language used.

1. Switches -- Poles and Throw. Switches used in motor control systems are designated as:

Simple toggle and push button switches (Fig. 3) with no overload protection are often used in control panels and for starting small motors.

3.   For many motors and large pieces of equipment a magnetic-type switch (Fig. 4) is used and has certain advantages not found in standard snap-action tumbler-type switches.

This type of switch is hand actuated, meaning that the operator pushes a button which causes the magnetic switch to close the circuit to start the motor. He pushes another button when he wishes the motor to stop.

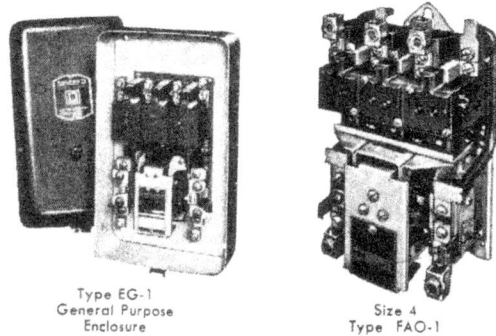

Type EG-1
General Purpose
Enclosure

Size 4
Type FAO-1

Fig. 4. Magnetic switches used to control large electrical loads and large motors.

The start button permits a small current to flow to an electromagnet which is a small coil of wire wrapped around an iron core. When this happens, the electromagnet attracts a lever or movable contact point, closing it onto a stationary contact point. The main leads to the motor are connected to these contact points. Thus, the motor current does not flow through the start button nor through the electromagnet.

An advantage of a magnetic switch is that it may be operated by pressing a button which may be located some distance from both the motor and magnetic switch. This leads to convenience and safety in starting and stopping a motor, especially if it is of large HP or if it must be controlled from one or more remote locations. A magnetic switch generally contains an overload device.

D.   Automatic control devices.

An automatic control system has two basic parts -- a sensing element which reacts automatically to certain conditions or changes in conditions, and a switch that is actuated by the sensing element to start or stop a motor or other devices.

The sensing element is much more accurate, positive, and dependable in its operation than is a human being. A timing element, for example, can automatically actuate a switch at 4:00 a.m. each day, whereas it

11

Fig. 10. SPDT Switches

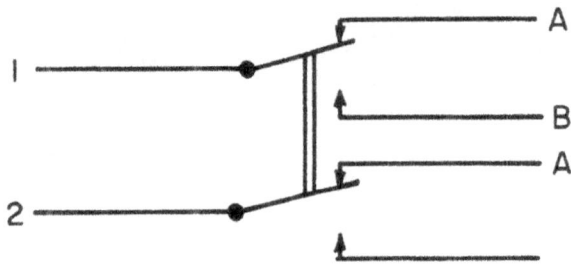

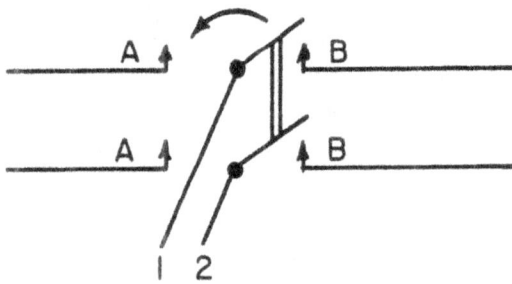

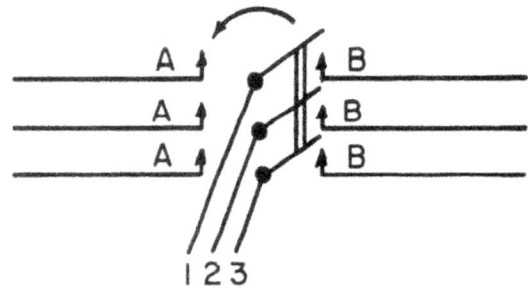

Fig. 11. DPDT Switches

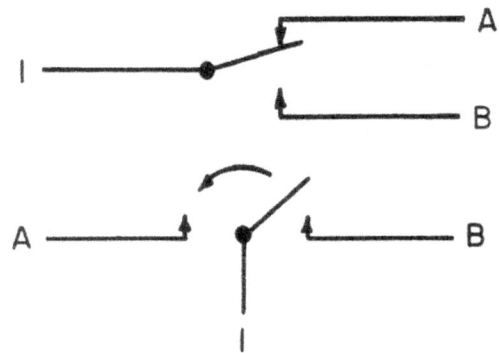

Fig. 12. 3PDT Switches

2. Switches -- Normally Open and Normally Closed - Hand operated switches can usually be left in either the closed (on) or open (off) positions. One exception to this is the ordinary doorbell button (Fig. 13). This is a normally open (N/O) switch. The force of your finger on the button is necessary to close the circuit and ring the door chime. When you release the button a spring causes the contacts to open. The circuit will remain open until the outside force acts upon the button again. Many of the switches in control systems are either Normally Open (N/O) like the doorbell button, or Normally Closed (N/C) like the switch shown in Fig. 14. This is the stop button used in magnetic starter circuits for motors. You will become familiar with this in the laboratory exercises.

14

a. Single-pole (SP) -
Single-pole switches
break only one of the
line wires and are
generally used with
120-volt circuits. A
snap-type, tumbler-
type switch used to
turn "on" the kitchen
lights is a single-pole
switch.

Fig. 5. Schematic Wiring Diagram
of Single-Pole (SP) Switch

b. Double-pole - Double-pole switches break two line wires and
should always be used on 240-volt systems (Fig. 6).

c. Three-pole - Three-pole switches are used with 3-phase circuits and in circuits where the switch is to control more than
one circuit (Fig. 7). They are also used in relays where a
holding circuit is required for pushbutton stations.

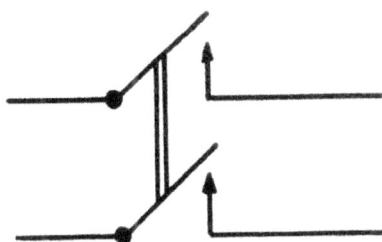

Fig. 6. Double-Pole
(DP) Switch

Fig. 7. Three-Pole Single
Throw Switch

If the switch has just
two positions -- ON
and OFF -- it is called
single-throw (ST). The
complete description of
the switch shown in
Fig. 8 is, therefore,
single-pole, single-
throw (SPST). The
switch shown in Fig.
9 is double-pole,
single-throw (DPST).

Fig. 8. Single-Pole Single
Throw (SPST) Switch

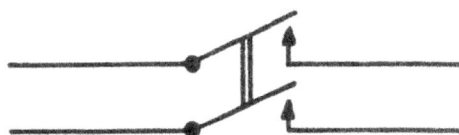

Fig. 9. Double-Pole Single
Throw (DPST) Switch

A double-throw switch
is one that switches
the line wire to one or
the other of two positions. Fig. 10 shows switches that are
single-pole, double-throw (SPDT). Fig. 11 shows double-pole
(DPDT) switches. Fig. 12 shows 3-pole, double-throw (3PDT)
switches.

13

SWITCHES AND SWITCH CONTROL CIRCUITS

This lesson plan is designed to cover two 1-hour periods.

TEACHING OBJECTIVES:

1. To familiarize the student with the characteristics of switches most commonly used in control circuits.

2. To teach the student how to connect a tumbler switch having a built-in overload protector into a motor circuit and to teach how the switch protects the motor from short circuit or overload damage.

3. To teach the student the function of a limit switch and how to connect such a switch into a motor circuit.

4. To teach the student how to connect 3-way and 4-way switches for the control of lighting circuits.

TEACHING AIDS:

1. Tools and Equipment

   Snap action tumbler switch - (1)
   Tumbler switch with overload protector - (1)
   Limit switch (Gen. Elec. Model CR115B201, or similar) - (1)
   Three-way swtiches - (2)
   Four-way switch - (1)
   Motor - (1) (May need extra motors for demonstration of overload
     protection)
   Socket and light bulb - (1)
   Connection wires

2. Visual Aids

   Blackboard
   Overhead projector and screen
   Transparencies:
       TP 1-S&S - Schematic diagram of SPST switch and motor; dia-
                  gram of switch with built-in overload protector;
                  diagram for demonstrating overload protection
       TP 2-S&S - Schematic diagrams of limit switch
       TP 3-S&S - Schematic diagram of three and four-way switches

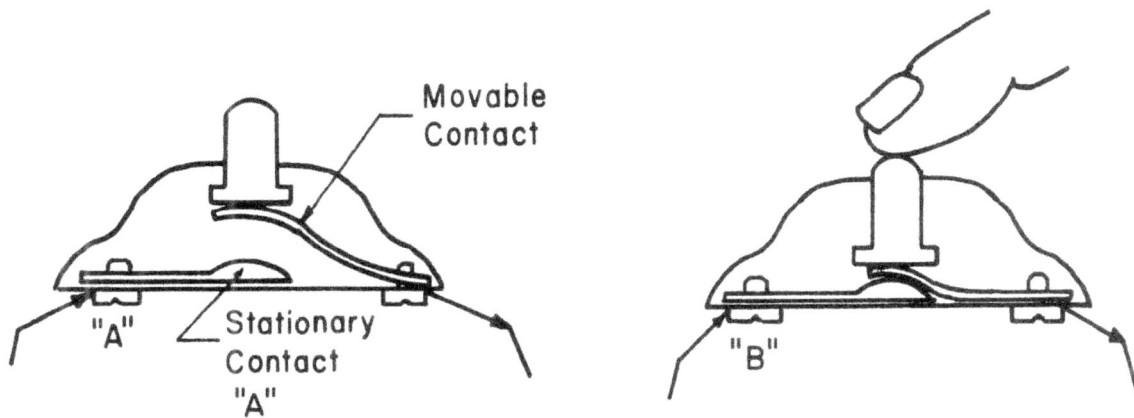

Fig. 13. Switch is normally open (N/O). "A" The movable contact acts as a spring to hold the two contacts apart. Current cannot flow. "B" Button is pressed and the two contacts touch, allowing current to flow.

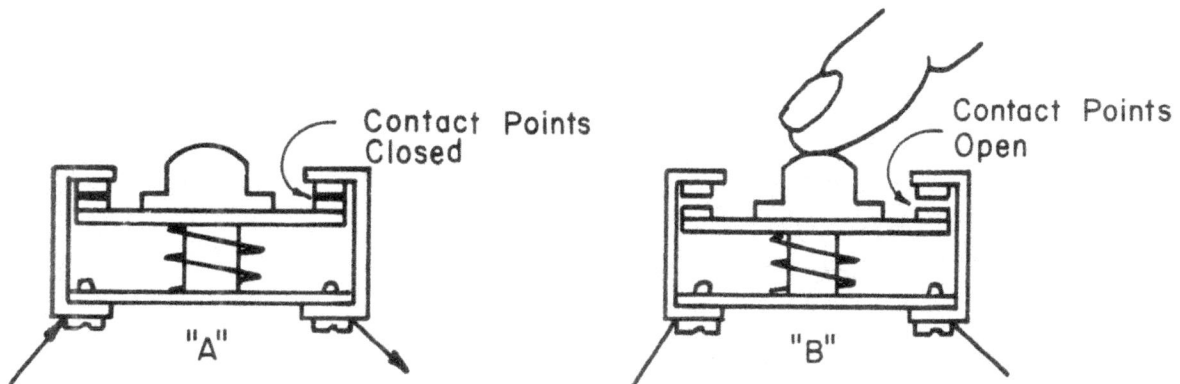

Fig. 14. Switch is normally closed (N/C). "A" Switch in closed position. The spring holds the contacts together so current can flow. "B" Button is pressed and contacts are opened so current cannot flow.

F. Exercises and skills

1. Have student make a list of controls termed as "switches".

2. Have students draw a wiring diagram of a simple control system consisting of one "switch" and one "load device." (Could be a simple lighting circuit).

3. Have the student draw schematic diagrams of SPST, SPDT, DPST, and DPDT switches.

b. How would you protect the motor against damage from overloads and short circuits when using this type switch as a motor control?

C. Using the tumbler switch with built-in overload protector:

1. Remove the cover plate and than take out the overload protector, which also is called a "heater." What is the ampere rating of the heater?

a. What would happen if more than the rated number of amperes went through the switch?

b. What is the maximum size 120 volt motor for which this control is adequate?

c. How do you select the proper size overload protector element?

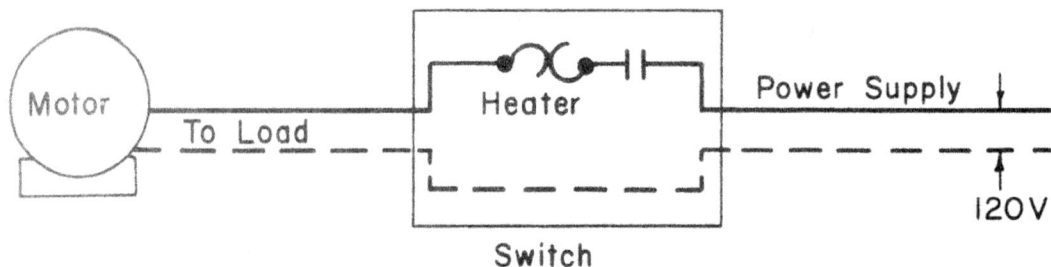

Switch

2. Connect the switch for controlling one motor. The wiring diagram is shown above. You will only need to connect the two line wires and the two motor leads at the ends of the switch.

3. Start and stop the motor several times.

4. Lift out the overload protector element. Will the motor run now that you have removed the element? If your answer is no, explain why.

5. To demonstrate operation of overload protection, connect a heavy load or several motors (3) in parallel to this switch.

a. How many amperes will flow through the switch when all 3 motors are running?

b. What should happen to the circuit and flow of current?

c. Did the overload protector open up and the motors stop?

d. Snap the switch to "off" and let the heater element cool.

e. Snap the switch to "on" and time the period necessary to cause the heater to open the circuit. What was the time?

18

3. Sources of Information

   Manufacturers' catalogs (Gen. Elec., Westinghouse, Leviton, etc.)
   Power suppliers
   Electrical contractors

PROCEDURE:

A. Have the students study each type switch listed above and answer the following questions about each switch: (See appendix for answers to questions)

   1. What is its ampere rating?

   2. What is the voltage rating?

   3. Can the switch be used to control a one HP motor that requires 16 amperes when connected to 120 volts? Can the switch also be used to control a 1 HP motor that requires 8 amperes if connected to 240 volts?

      What would happen to the switch if it was connected to a motor that required more amperes than the rated capacity?

   4. Is the switch single pole or double pole?

      Can the switch be used as either a normally open or normally closed switch?

B. Using the tumbler or toggle switches:

   1. Connect the switch so that you can control one motor (or a socket and light bulb) connected to 120 volts. Do you connect the grounded (neutral) wire into the switch or do you connect the ungrounded (hot) wire? Why?

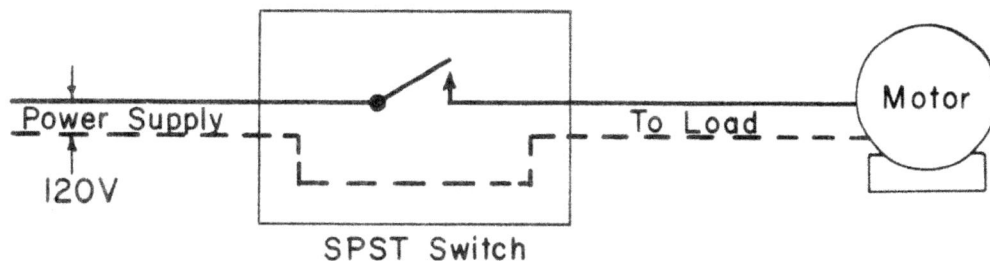

SPST Switch

   2. Connect the switch and start and stop the motor several times.

      a. Does this switch provide short circuit and overload protection for the motor?

17

a.   Energize the circuit. Does the motor start and run?

b.   Press on the lever arm roller. Does the motor start and run?

c.   Is this connection normally open or is it normally closed?

2.   Follow the diagram below, make the connections, and perform the work described.

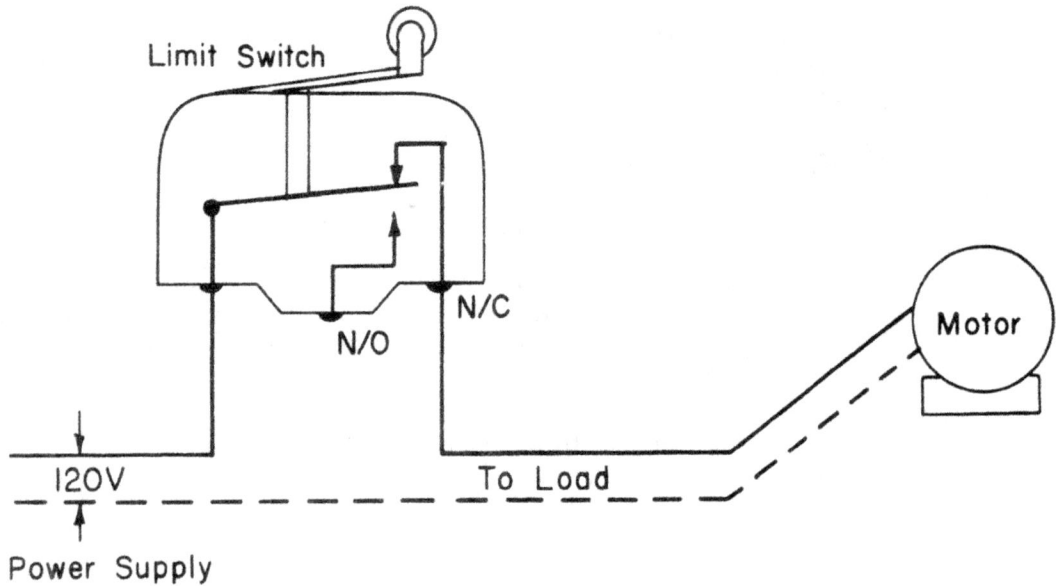

a.   Energize the circuit. Does the motor start and run?

b.   Press on the lever arm roller. Does the motor stop?

c.   Is this connection normally open or is it normally closed?

3.   Follow the diagram below, make the connections, and perform the work described.

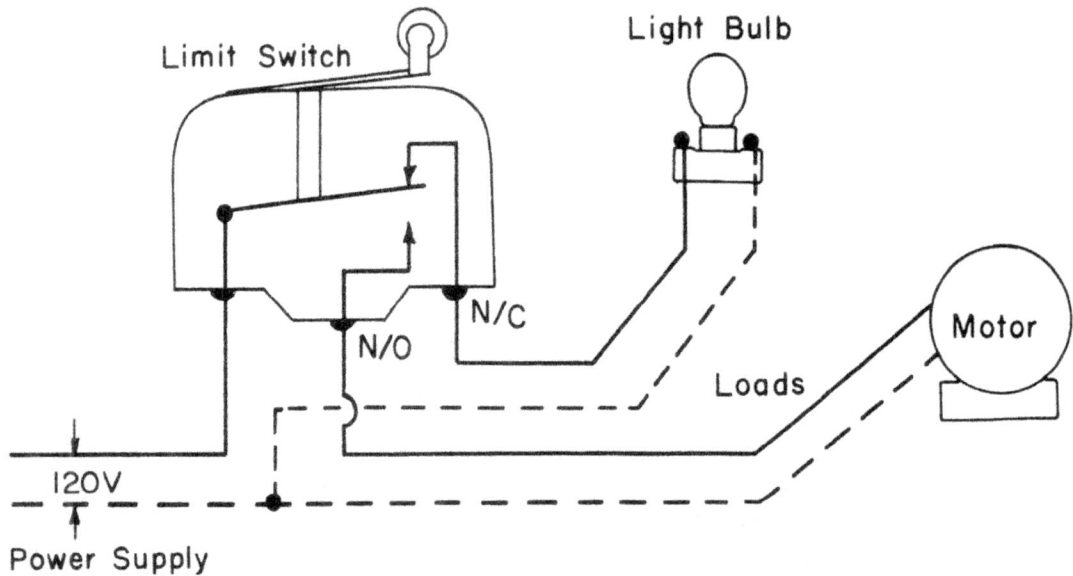

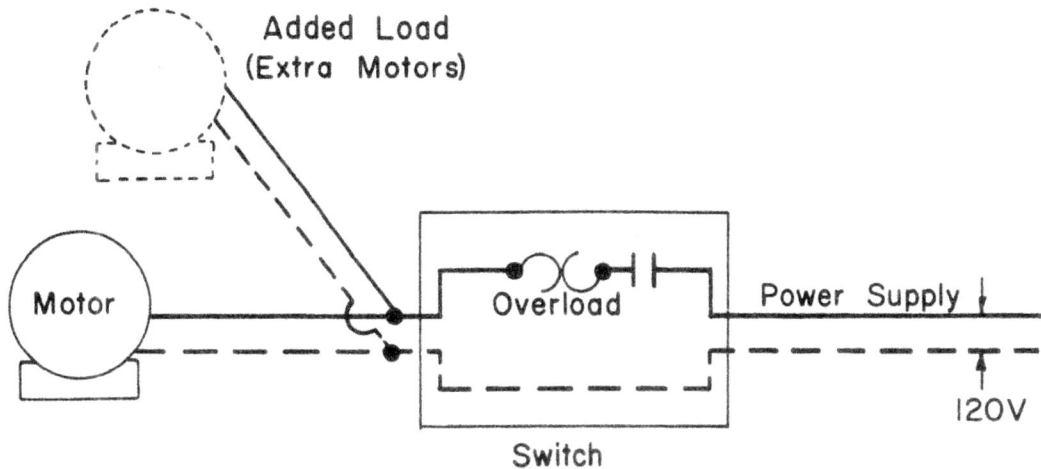

DIAGRAM SHOWING CONNECTIONS FOR DEMONSTRATING
OVERLOAD PROTECTION

6. How do you reset the overload protector?

In actual practice, the connection of 3 motors through this one switch would not be done. The major advantage of a switch with overload protection is that the heater can be sized to protect one motor. If 3 motors were connected, the heater would have to be so large that none of the motors would receive adequate overload protection. Three motors are connected in this exercise only to show that the total amperes will cause the switch to open.

D. Discuss the limit switch with students, indicating function of this type switch. Make a list of applications.

1. Follow the diagram below, make the connections shown and demonstrate the action of the switch.

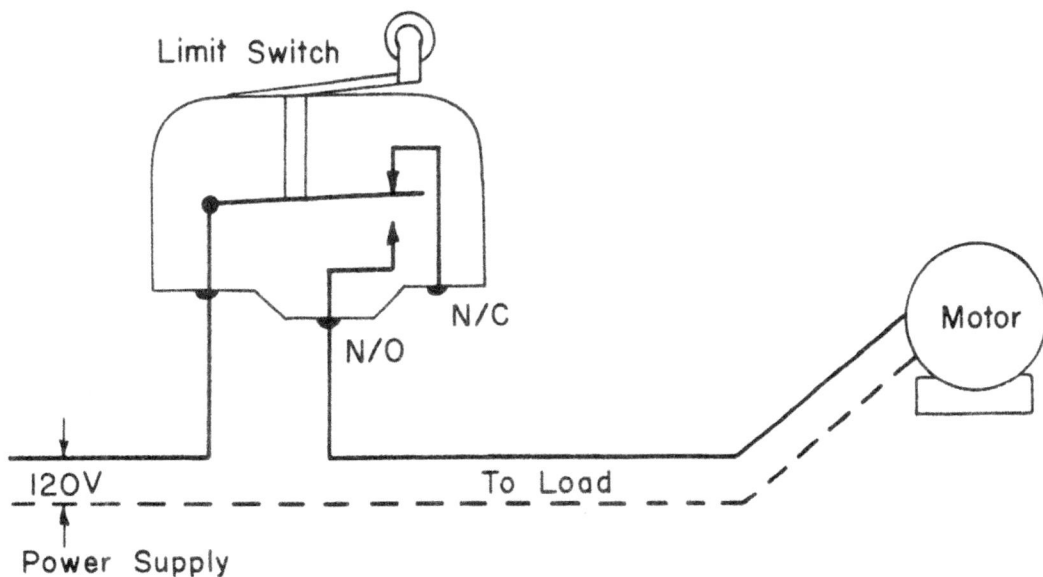

19

7. Make the connections as shown in the diagram below.

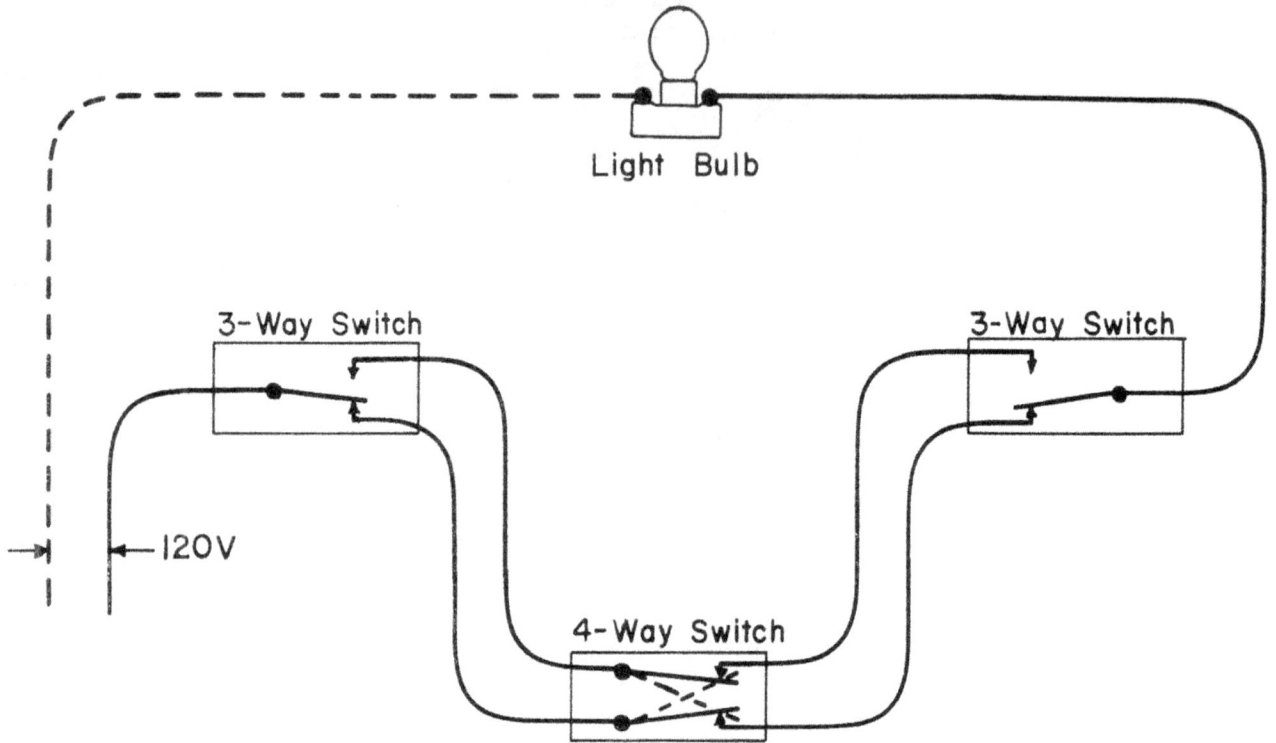

Light Bulb

3-Way Switch

3-Way Switch

←—120V

4-Way Switch

8. How many terminals does the 4-way switch have?

9. Draw the above diagram on the blackboard, draw lines on the 3-way and 4-way switches making connections so that the light will burn.

10. Draw a diagram showing how two 3-way and two 4-way switches can operate a light from four different places.

a. Energize the circuit. Does the motor start and run or is the light bulb on?

b. Depress the roller arm. Describe the effects on the motor and on the light bulb.

E. Using the 3-way and 4-way switches, perform the demonstrations indicated and discuss the questions asked.

1. Could these switches be used to control large motor loads?

2. What differences do you see between these switches and the others you have studied?

A 3-way switch has three terminals instead of only two, such as are used on a single pole switch.

3. Connect the three-way switches so that you can control the light from two different points.

4. Do the words "on" and "off" appear on the 3-way switches? Why?

5. Does power travel through one or both switches before reaching the lamp?

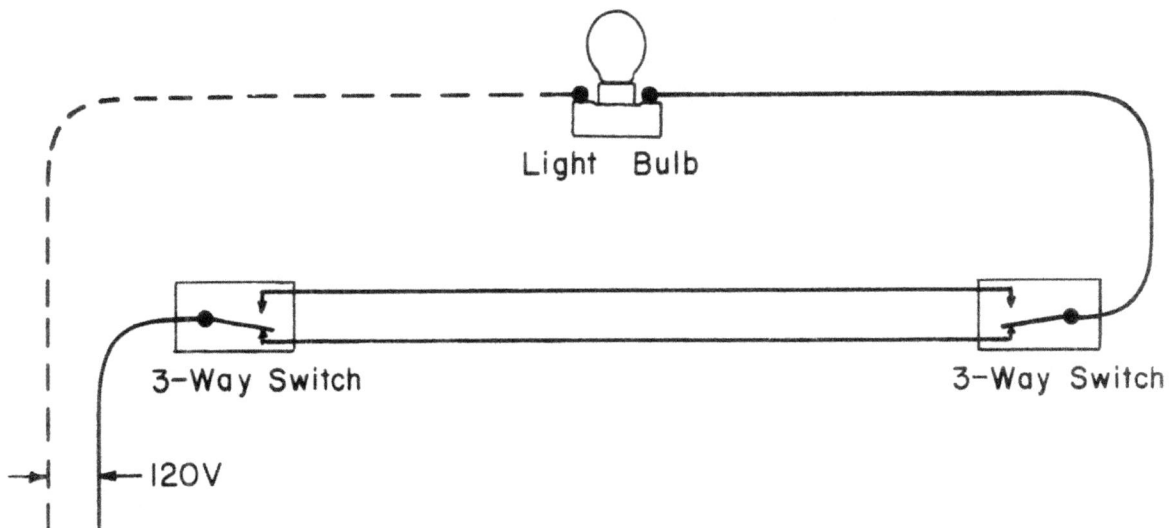

Light Bulb

3-Way Switch   3-Way Switch

←120V

6. Trace the circuit from the power source through the switches and the lamp.

21

main load is handled through the switch points that are closed when the electromagnet attracts the iron lever which is connected to them.

1. Purpose - Relays have one general purpose -- to control a circuit carrying a large current by means of a low-current circuit. For example, relays are usually used in photoelectric circuits. The weak current produced by the photoelectric cell cannot directly operate devices. such as motors. The small current can, however, operate a relay. The relay then acts as a switch to turn "on" or "off" the main power circuit which operates the heavier load. Relays are the heart of any automatic equipment and make automation possible.

2. Components - The basic parts of a relay are the solenoid, or electromagnet, made up of a coil and a movable armature or iron core and a set of contacts.

    Electrically, the relay has two circuits:

    a. The control circuit (which may often, but not necessarily always, be of a lower or different voltage).

    b. The load circuit which has a set of contacts that are opened or closed by the control circuit.

3. Operation - From the diagram in Fig. 1, you notice that there is a small coil of many turns of fine wire. When an electric current flows in the coil, electromagnetism is produced. This attracts the movable arm, pulling it down until it makes contact with another contact point. The main line current then flows through the movable arm and the contact points to the load. The relay shown in Fig. 1 is a SPST normally open type.

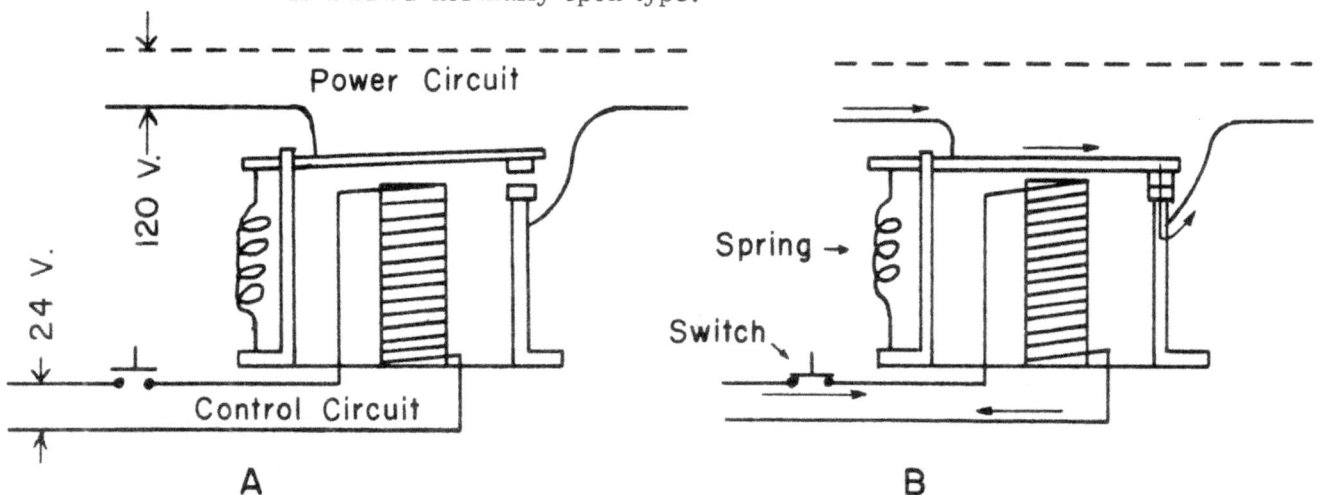

Fig. 1. Schematic diagrams of magnetic relay. A. Push-button switch is open so no current flows in control circuit. Power circuit is also open. B. Push-button switch is closed so current flows in control circuit. Electromagnet closes switch in power circuit so power current flows.

RELAY DEVICES

This lesson plan is designed to cover two 1-hour periods.

TEACHING OBJECTIVES:

1. To familiarize the student with the characteristics of relays and to help the student understand the function of relay devices in control circuits.

2. To teach the student how to connect a relay into a control system.

3. To teach the student how to connect a time delay relay into a motor control circuit.

TEACHING AIDS:

1. Tools and Equipment

    SPST Relay - (1)
    DPST Relay (Potter & Brumfield - Type PR7AY or equivalent - (1)
    Time delay relay (115 NO 10 - Amperite) - (1)
    8 point tube base for time delay relay - (1)
    Motor - (1)
    Switch control device (thermostat - photoelectric cell, etc.) - (1)
    Connection wires

2. Visual Aids

    Overhead projector and screen
    Transparencies:

    TP 1-R - Schematic diagram of magnetic relay; schematic diagram of SPDT relay
    TP 2-R - Schematic diagrams of DPST relay and DPST relay interconnected with time-delay relay
    TP 3-R - Schematic diagrams of SPST switch with time-delay relay; SPST relay interconnected with thermostat

3. Sources of Information

    Manufacturers' catalogs (Square D, Westinghouse, Allen-Bradley, Gen. Elec., etc.)
    Myers Prefiled Agricultural Catalogs
    Control instruction sheets
    Power suppliers
    Electrical contractors

PROCEDURE:

A. Magnetic relays are very useful in controlling relatively large electrical loads with small sensing elements. The use of relays is much like the use of power steering or power brakes on an automobile. It requires only a small force to turn the wheel or push the brake pedal because the operator is merely actuating the real driving mechanism by remote control. Similarly, the control circuit of a relay is a small electrical load - that which is required by the electromagnet only. The

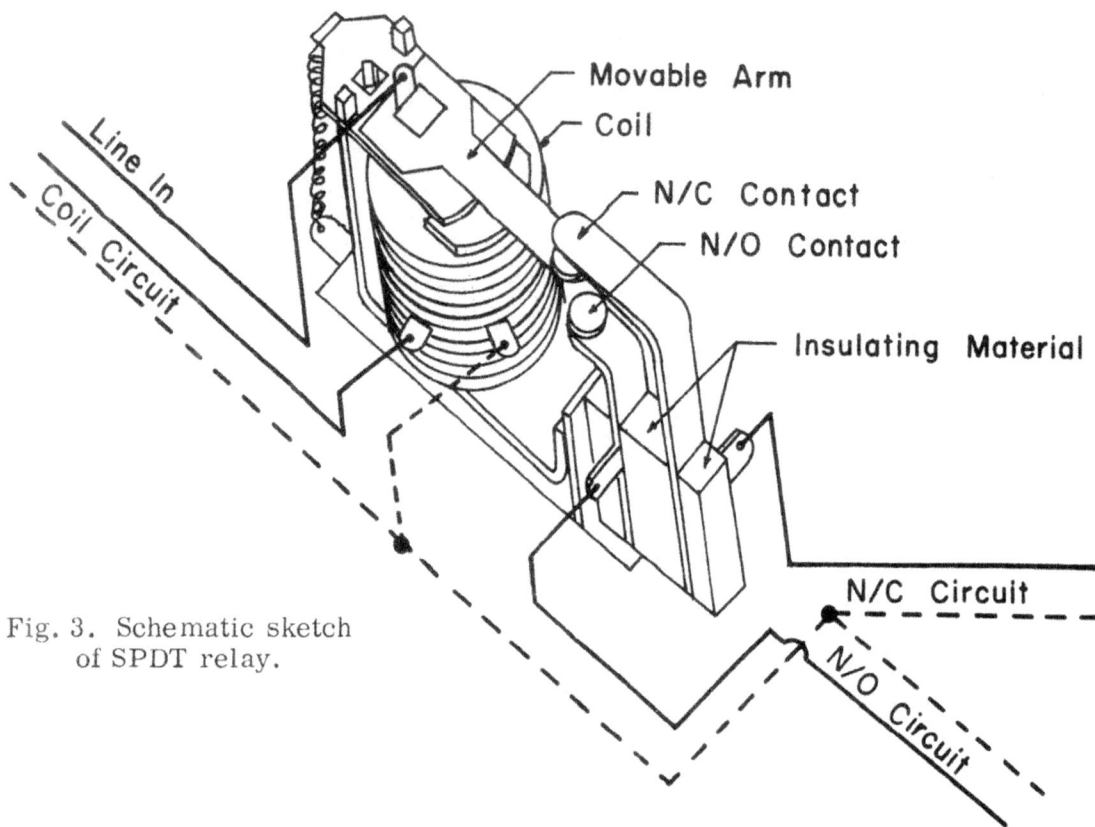

Fig. 3. Schematic sketch
of SPDT relay.

c. A double pole-single throw (DPST) relay has two sets of load contacts and, like the SPST relay, the contacts may be either normally open or normally closed. The DPST relay is ordinarily used to control 240 volt loads by opening and closing both hot line conductors. See Fig. 4.

When the coil is not energized, the contacts on the movable arm are separated from the stationary contacts. Therefore, a normally open (N/O) switch exists. When the coil is energized, the movable arm closes the contacts. Current can then flow from each hot line to the load.

B. Study of Relays - Have the students study each type relay available and answer the following questions about each piece of equipment:

1. What is the ampere rating?

2. What is the voltage rating?

    a. Voltage of coil

    b. Voltage of relay load circuit

3. What is the largest size load that can be controlled?

    a. Load in watts or kilowatts

    b. Size motor

4. Determine the number of poles or contacts on each relay and whether it is a normally open or normally closed relay.

    a. A single pole-single throw (SPST) relay (Fig. 2) may have either normally open or normally closed contacts. The contacts are usually held in position by some mechanical means such as a spring. In a normally open relay, the spring holds the contacts open until the coil is energized and the contacts are closed by the action of the coil and armature. In a normally closed relay, the action is reversed.

Fig. 2. Single Pole-Single Throw Relay

    b. A single pole-double throw (SPDT) relay has two sets of contacts, one is normally open and the other is normally closed. This relay may function as a SPDT switch. It may control two separate devices which must not operate simultaneously.

        Fig. 3 is a schematic sketch of a SPDT relay. Many farmstead-type relays use 120 volts for the coil energizing circuit. In the normal position, the movable arm touches the upper, or N/C, contact. If a line wire was connected as shown, current would flow from the line wire through the movable arm, to the N/C contact and on to the N/C load circuit. When the coil circuit is energized, the electromagnet would pull the movable arm downward so that it touches the N/O contact. Current would then flow from the line through the movable arm, to the N/O contact and on to the N/O load circuit.

D. Exercises and skills.

1. Connect the DPST relay and motor by following the diagram below.

DPST Relay

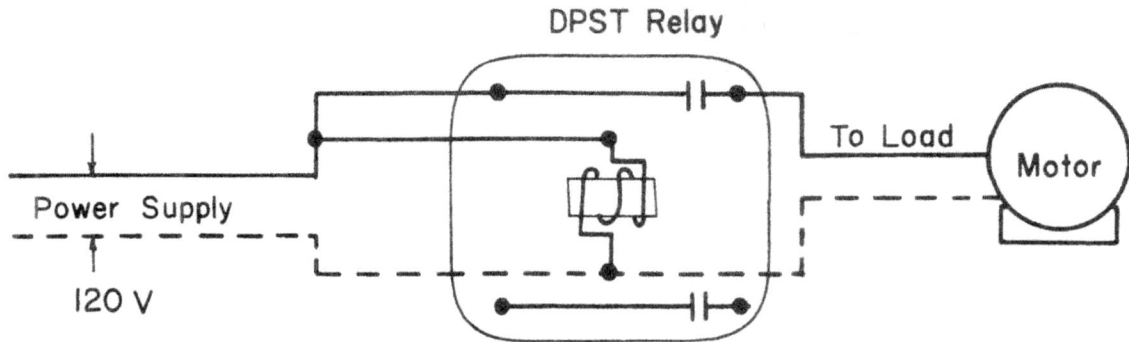

2. Energize the circuit. Does the motor start and run?

   a. Why is only one contact used?

   b. De-energize the circuit. Change the connections on the relay so that current flows to the motor through the other contact.

3. Perform the following additional work:

   a. Connect the thermostat controller, the SPST relay, and a motor by following the diagram below.

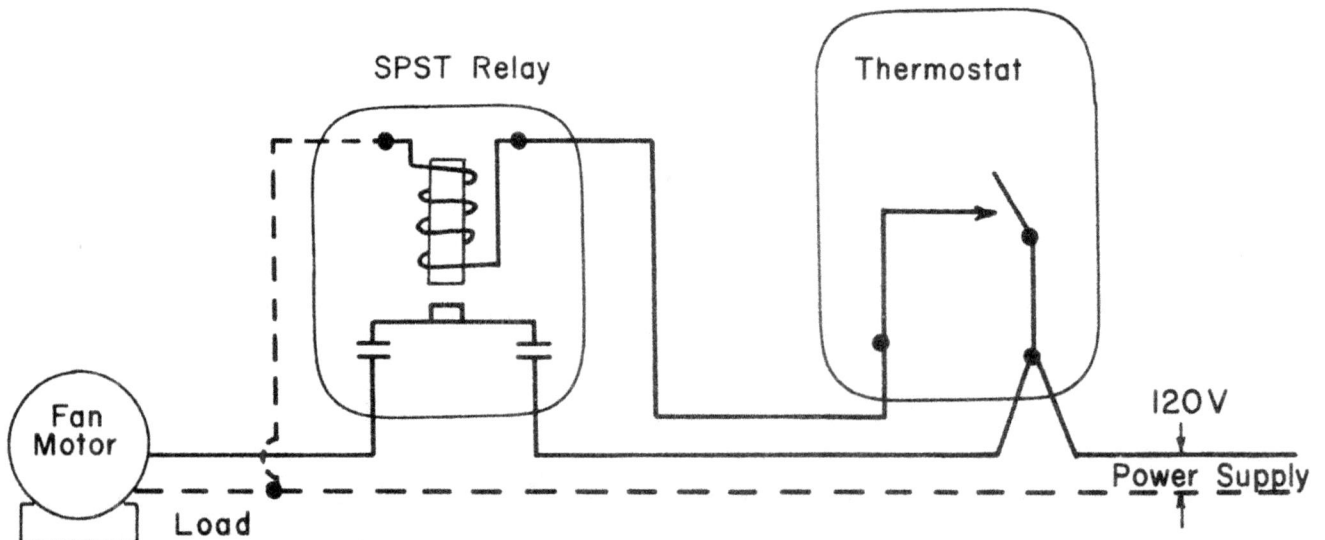

   b. Repeat the same work sequence outlined above.

   c. Why is a relay (or magnetic starter) used with the thermostat controller when a motor is connected?

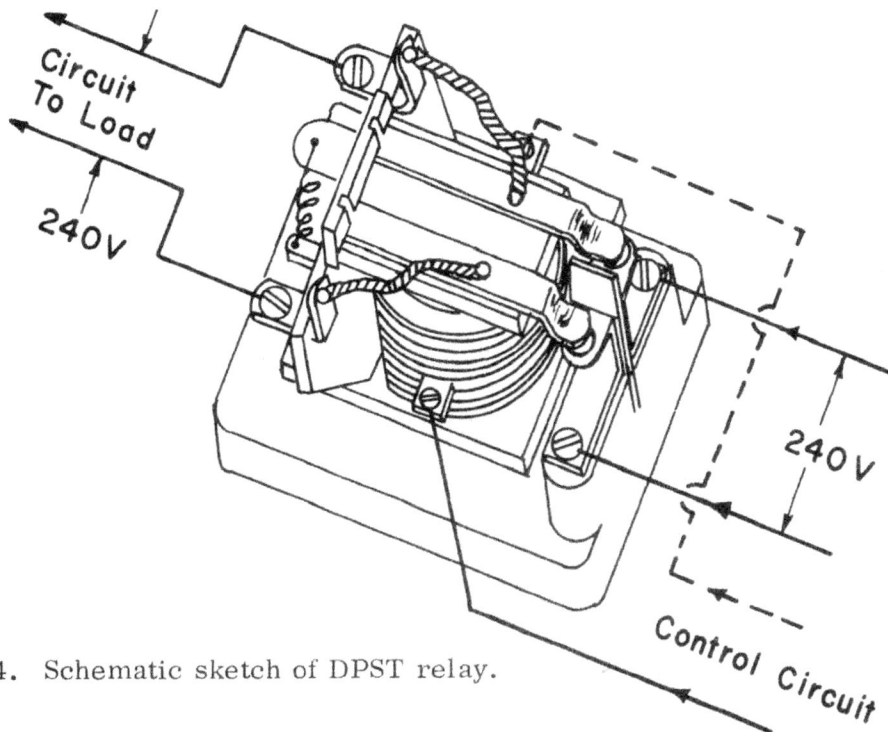

Fig. 4. Schematic sketch of DPST relay.

C. Study of relay circuits.

1. The components of a relay control system consist of a power source, a switch or automatic sensing device which operates the control circuit, the relay device, and the load which is connected to the load circuit.

2. Load Circuit - The electrical circuit to the motors or equipment being controlled is called the load circuit. This circuit is opened or closed by the contacts of the relay. The contacts are constructed so that large amounts of current may be conducted from the power source to the load. This current may be started or stopped by closing or opening the relay contacts.

3. The control circuit consists of switches or sensing devices and the relay coil. Since the load to be switched is carried by the load circuit, the only current flowing in the control circuit is that required to energize the coil. This is usually quite small (1 ampere or less). Because of this, the control circuit is often operated at low voltages, and the relay is easily adapted to remote control systems by the use of small control wires.

The control circuit may contain any number of manually operated switches or any number of sensing devices, in any combination, which are capable of opening or closing a circuit.

27

(2) Energize the circuit. Does the bulb come on immediately? Wait a few seconds. Does the bulb go on or off?

(3) De-energize the circuit. Does the bulb go on or off immediately? Why? Wait 10 seconds or so and repeat step 2.

(4) The contact points are rated at 3 amperes. Can you connect this time delay relay directly to a 1/4 HP motor?

c. Read the following explanation and perform the work described:

(1) Because a 1/4 HP or larger motor requires a greater current than the 3 ampere rating of the contact points of the time delay relay, an additional relay (magnetic starter) must be used with such a motor. The time delay relay then connects only to the coil of the additional relay and the motor connection is made through the additional relay contacts.

(2) Connect the motor to the 115 N/O 10 time delay relay by following the diagram below.

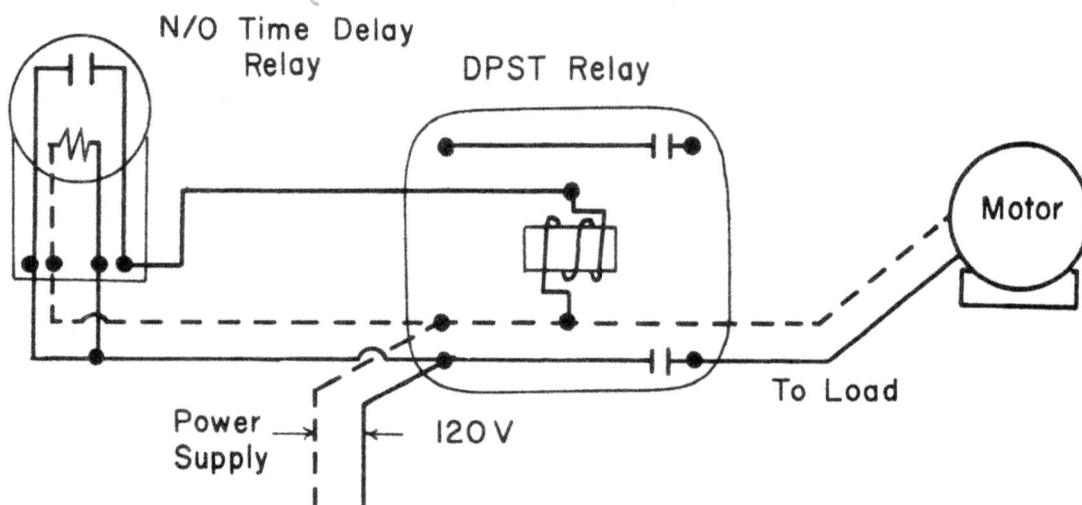

(3) Energize the circuit. Does the motor come on immediately? Wait about 10 seconds, does the motor then come on? Explain the action that occurs. De-energize the circuit. Time the delay in motor action.

d.   Can you think of a good farm application for using a thermostat controller?

E.   Time delay relays

1.   Purpose - A time delay relay is a device which delays for a predetermined time the opening or closing of a switch.

2.   Operation - The time delay relay consists of a heater element which actuates a bimetallic strip. When current is supplied, the heater causes warping of the bimetallic strip which closes a set of contacts. When the power is turned off, the heater goes off and the bimetallic strip returns to its original position opening the circuit.

3.   Application in control circuits:

a.   Examine the 115 N/O 10 time delay relay. The type used in this exercise has the physical appearance of a radio or TV tube. The 115 in the type designation 115 N/O 10 means that the heater element in the relay is designed for 115 volts. The N/O means that the relay contacts are normally open. The 10 means that the contact points will close 10 seconds after the heater is energized. When the heater is energized with a flow of current, a 10 second period is required before the heat will cause the metal plate to bend and thus close the contacts. This closes the switch. Locate the two contact points. Are they together or are they apart?

b.   Perform the following work:

(1)   Connect the 115 N/O 10 time delay relay by following the diagram below. Note that the switch within the relay is in series with the light bulb, therefore the full ampere flow goes through the switch. The contact points are rated at 3 amps. What is the maximum size bulb that can be used directly with this relay?

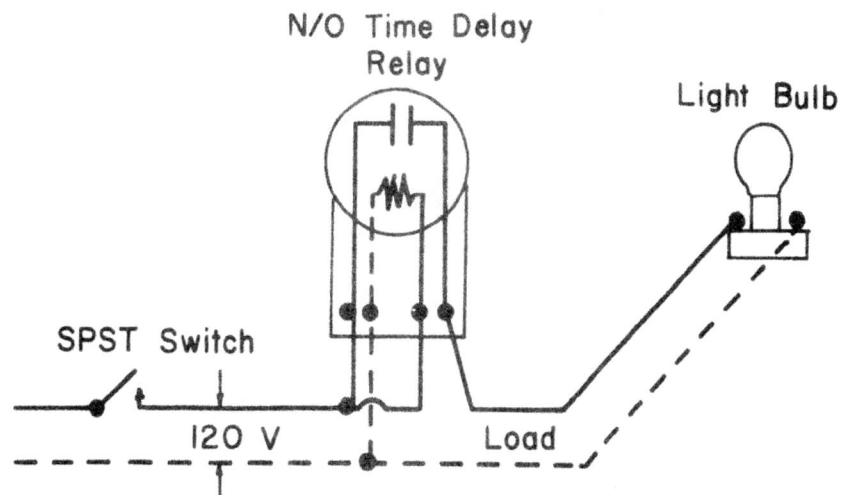

N/O Time Delay Relay

Light Bulb

SPST Switch

120 V

Load

larger) and to provide overcurrent protection for the motor. Motor controllers can also be used for connecting large heating and lighting loads to the source of power. The overcurrent protection is not used when the device is used for this purpose.

Motor controllers may be either manually or magnetically operated. The controllers vary greatly in size and are built to control various size and voltage motors.

B. Manual controller - The manual motor controller is operated by pushing a "start" or "stop" button which is built into the controller. (Fig. 1). It may be operated from only one position and may not be controlled remotely or automatically. The motor is started at full line voltage and the overload protection for the motor is provided by reliable thermal type overload relay heaters.

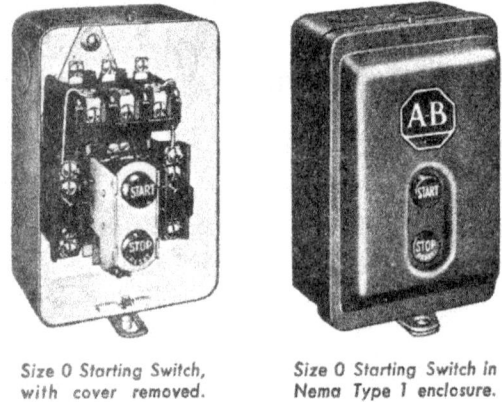

Size 0 Starting Switch, with cover removed.

Size 0 Starting Switch in Nema Type 1 enclosure.

Fig. 1. Manual Motor Controller

C. Magnetic motor controller

1. Components - The magnetic motor controller consists of a magnetic contactor and an overcurrent device in a common enclosure. See Fig. 2. The contactor is simply a large relay generally rated at 25 or more amperes. The contacts are opened or closed by the magnetic action of a coil and movable core operated by a control circuit.

Size 00

Size 0

Fig. 2. Magnetic Motor Controller

MOTOR CONTROL DEVICES

This lesson plan is designed to be covered in a 1-hour period.

TEACHING OBJECTIVES:

1. To familiarize the student with the characteristics and types of motor control devices and to help the student understand the function of a motor control device in an electrical circuit.

2. To teach the student the function of each component part of a motor control device and the identification of the component parts in a schematic wiring diagram.

3. To teach the student how to connect a commercial type magnetic starter switch and start-stop pushbutton stations into a motor circuit.

TEACHING AIDS:

1. Tools and Equipment

   Magnetic starter switch (Square D Model 8536-BG1, or equivalent) - (1)
   Start-stop pushbutton stations (Square D Type B-30) - (2)
   Electric motor - (1)
   Connection wires

2. Visual Aids

   Overhead projector and screen transparencies:
   TP 1-M - Schematic diagram of motor contacts; diagram of motor contacts and magnetic coil; diagram of motor contacts, magnetic coil and overload protection; diagram of start and stop station
   TP 2-M - Schematic diagram of complete motor starter; diagram of motor starter and start-stop pushbutton stations
   TP 3-M - Schematic diagram of motor starter and two start-stop pushbutton stations

3. Sources of Information

   Manufacturers' catalogs (Square D, Westinghouse, Allen-Bradley, Gen. Elec., etc.)
   Wiring diagrams and instructions for starter
   Power suppliers
   Electrical contractors

PROCEDURE:

A. Purpose - The basic function of a motor control device or magnetic starting switch is to start and stop electric motors (1 horsepower and

(c) To give current over-load protection to the circuit, an overload protector (or heater) is added and is connected to the hot conductor. Terminal T_1 then is extended to one side of the overload heater unit.

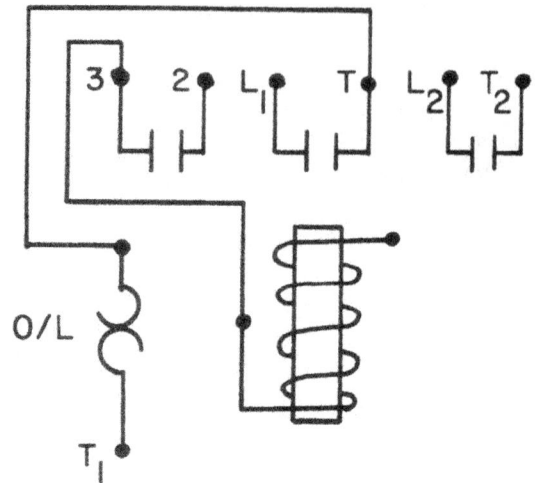

(d) If this overload protector is used with no other safety mechanism, the motor would start automatically as soon as the heater cooled. This is undesirable. To give added protection, a small N/C switch is added within the overload protector unit. If an overload occurs, this small switch opens. Current cannot then flow through the coil and the 3-pole switch opens, stopping the flow of all current. Additional wiring is added for the controlling device (pushbutton station), and, for easy connections in the laboratory exercises, the wiring is extended to separate terminals on terminal blocks.

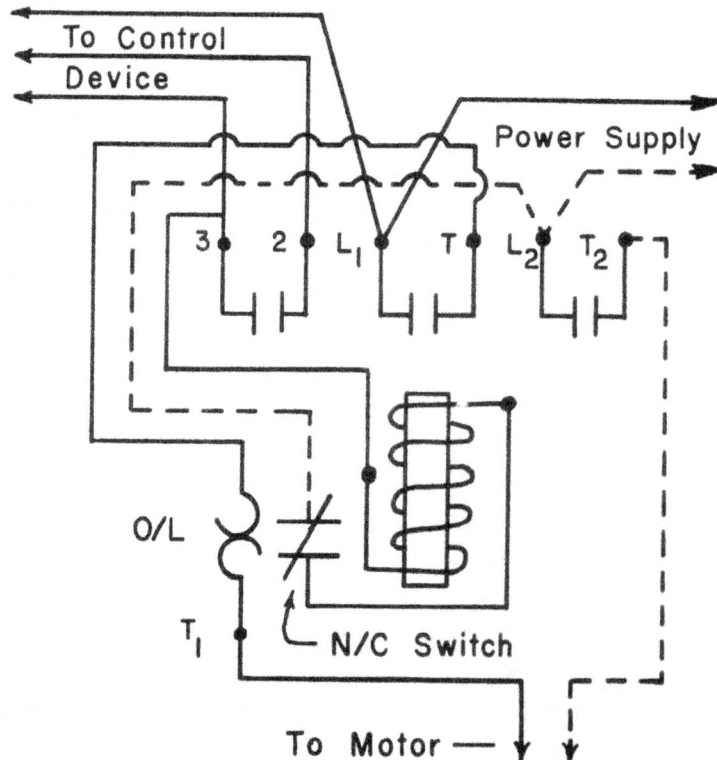

2. Rating - The contacts are rated in volts, amperes, and horsepower. The control circuit may be of the same or lower voltage.

3. Overload Device - The overload device is sized according to the motor or load it controls and will open a contact in the control circuit if an overload occurs. Before the motor can be started again, the overload device must be manually reset.

D. Study of magnetic starter and schematic diagrams

1. Remove the cover of the magnetic starter and note the parts of the switch. Make sure that you can identify the contact terminals, coil, magnet assembly, overload relay, relay heater, and reset mechanism.

2. Is this a single-pole, a double-pole, or is it a three-pole switch?

3. By studying the nameplate on the magnet assembly, state the maximum size motor connected to 120 volts for which this switch is adequate. What is the maximum size motor connected to 240 volts for which this switch is adequate?

4. Does the switch have an overload protector? Where is it? Is the overload protector adjustable? How do you select the proper size heater unit for use in the protector?

5. On the switch itself, locate and point out the terminals L_1, L_2, T_1, T_2, 1, 2, 3; the magnet assembly, the coil, the overload protector unit, and the heater unit.

(a) Inside the magnetic starter are 3 separate poles, each a N/O switch. The ungrounded (hot) conductor connects to L_1. The grounded (neutral) conductor connects to L_2. The other switch is used for the controlling device circuit and is between terminals 3 and 2.

(b) A magnetic coil is added to the 3 pole switch. The coil is composed of many turns of fine wire wrapped so as to fit around a soft iron core. When current flows through the coil, a magnet is created.

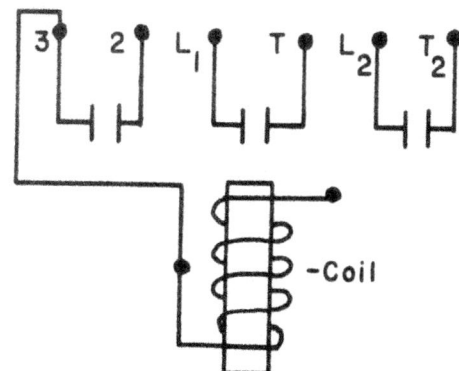

33

One of the advantages of a magnetic starter is that any number of start-stop stations may be placed in the circuit so that the motor may be controlled from several places. All of the start buttons in such an arrangement are placed in parallel with each other. This allows current to flow if any one of the buttons is pressed.

The stop buttons are all placed in series. Current flows continuously through the stop circuit until one button is pressed, opening the circuit and stopping the flow of current to the coil of the magnetic starter.

To simplify the wiring for this operation, one terminal of the start button and one terminal of the stop button in each station are connected together.

E. Pushbutton motor control stations

  1. Purpose - The momentary-contact, manually-operated pushbutton station, when used with a magnetic controller, is the most common type of remote-control device used for starting and stopping electric motors. It may also be used in other type relay control circuits. This type pushbutton control cannot be used as a switch by itself but requires other equipment, such as a motor starter or relay device in performing its control function.

  2. Current rating - The pushbuttons are usually made of Bakelite or similar molded insulation, and the contacts are silver plated. The current rating of the contacts is small (1, 2, or 3 amps) since the motor or load current does not flow through the pushbutton contacts.

  3. Study of pushbutton stations and schematic wiring diagrams.

    a. Remove the cover of the station and locate the contact points under the stop-button and the contact points under the start-button.

    b. Push the start-button. Do the contact points open or do they close? Is the start-button N/O or is it N/C?

    c. Push the stop-button. Do the contact points open or do they close? Is the start-button N/O or is it N/C?

    d. What is the maximum voltage on which this push-button may operate? Is it suitable for the voltages you have at home? What voltages do you have at home?

    e. The start side of the station has 2 terminals and the stop side also has 2 terminals. Only 3 wires, however, are brought outside. Why aren't there 4 wires brought outside?

The following diagrams illustrate the action on a start-stop station:

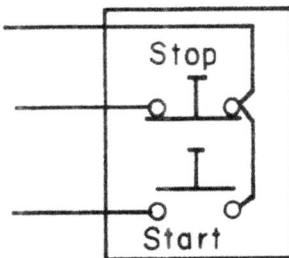

Diagram showing normal position.

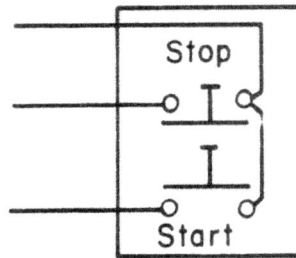

Diagram showing Stop button pressed.

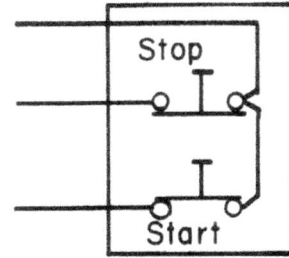

Diagram showing Start button pressed.

2. Connect 2 start-stop stations to the magnetic starter and control the motor from each station.

    a. Follow the diagram below to make the connections. Start and stop the motor several times.

    b. Does the current to the motor flow through the start-stop stations? If not, what current does flow through the stations?

    c. How does this current flow characteristic simplify the wiring requirements?

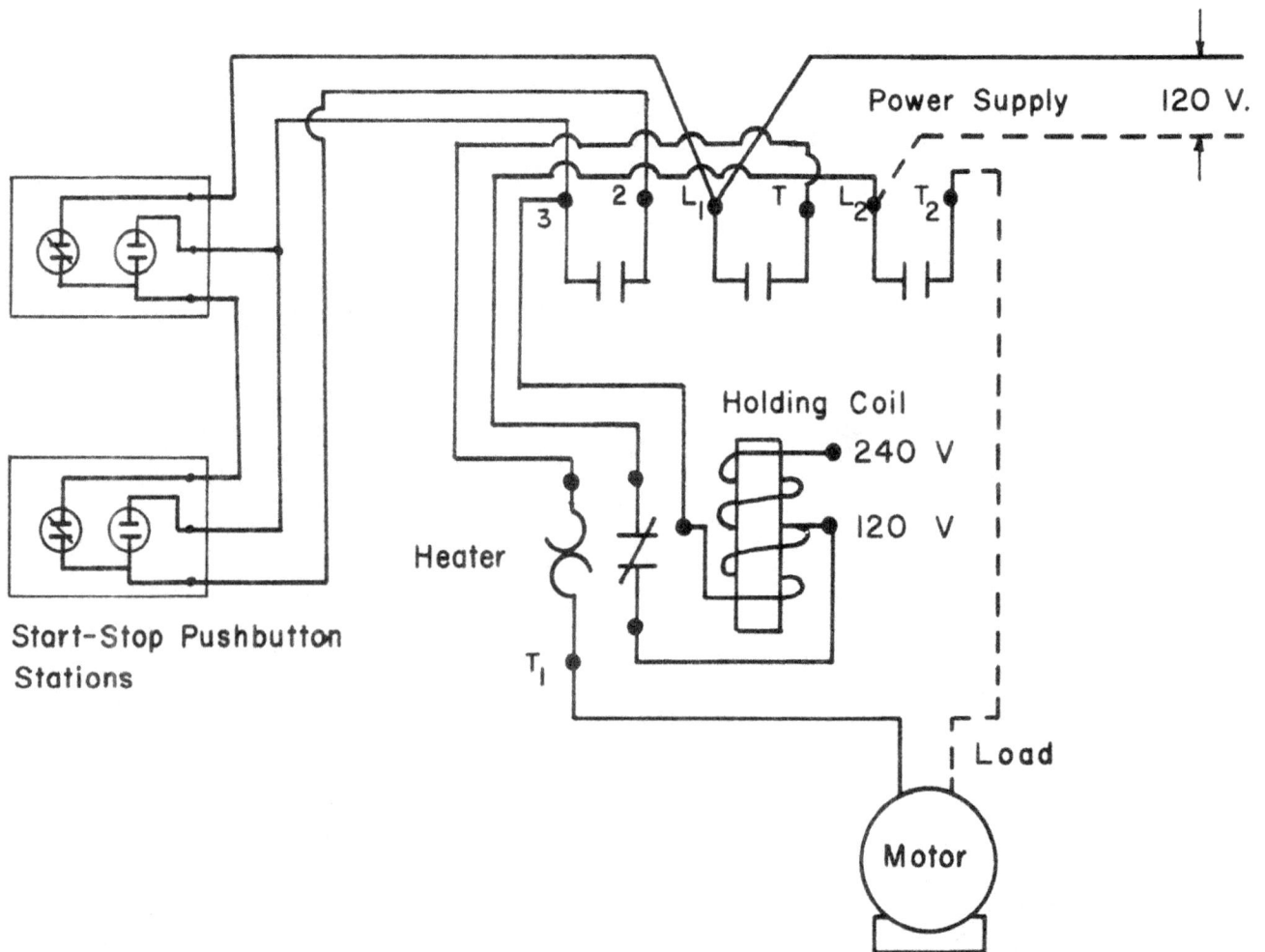

    d. Would the above control mechanism protect the motor if it became overloaded? How?

F. Exercises and Skills

1. Study the diagram below and make the connections as shown. Use the magnetic starter and one start-stop station.

   a. Energize the circuit and push the START button. Does the motor start and run? Did the magnet assembly move? Release the START button. Does the motor continue running?

   b. Push the STOP button. Did the magnet assembly move? Did the motor stop?

   c. Start and stop the motor several times and notice the action of the magnetic starter.

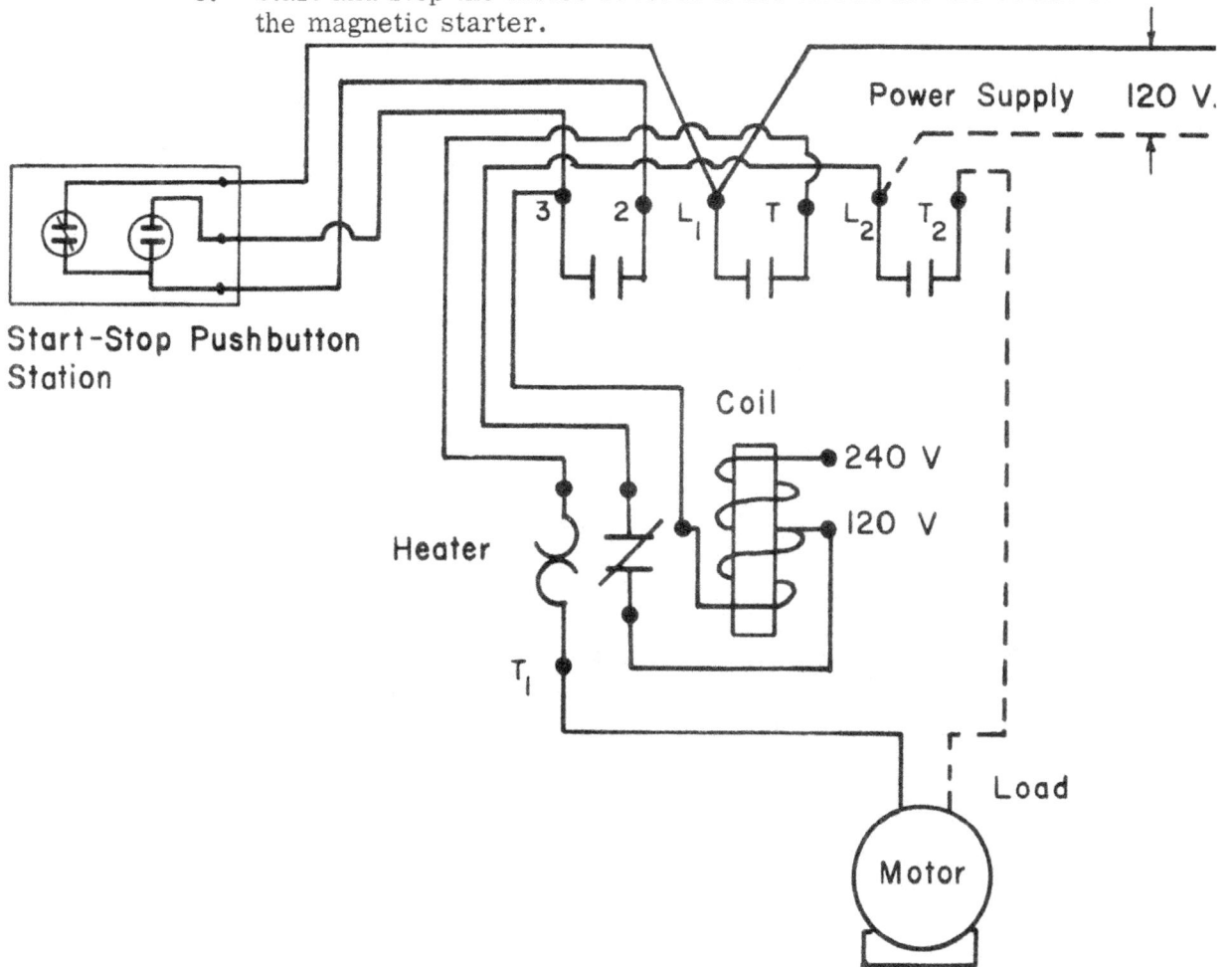

When the start button is pressed, current may flow from L_1 through the stop button which is N/C, on through the start button, on through 3 to the magnetic coil. When the coil is energized, the 3 switches within the magnetic starter close and current may flow from L_1 to T_1 to the motor. The neutral goes from L_2 through the closed switch to T_2, and then to the motor.

To stop the motor, the coil needs to be de-energized. Pressing the stop button will open the circuit to the coil. The 3 switches within the magnetic starter open and the motor will stop.

3. Sources of Information
Manufactuter
Manufacturers' catalogs (Paragon, Minneapolis-Honeywell, Intermatic,
   White-Rogers, etc.)
Instruction sheets and wiring diagrams for controls
Myers Prefiled Agricultural Catalogs
Power Suppliers

PROCEDURE:

A. Purpose - An automatic sensing device is a piece of equipment which is
capable of sensing a change, such as time, temperature, humidity, light,
pressure, etc., and then mechanically actuating an electrical contact or
switch. The device may be set to react to a specific condition within
its range of sensitivity. The automatic sensing device may be designed
to be operated on line voltage or on low voltage circuits. These devices
are rated in both volts and amperes. A line voltage control has heavier
contacts and is generally not as sensitive as a low voltage device.

B. Temperature controllers

1. Temperature sensing control devices - Temperature control devices are commonly called thermostats. They operate on three
basic principles. Each type responds to pre-set temperature
ranges by opening or closing a contact switch.

a. Bimetallic thermostat - This device operates on the principle
that different metals expand or contract at different rates due
to changes in temperature. Two dissimilar metal strips are
bonded together. When the temperature changes, the bimetallic
strip warps and this action closes or opens an electrical contact.

b. Liquid-gas thermostat - A liquid with a low boiling point is
enclosed in a wafer or bellows connected mechanically to a
switch which opens or closes a contact when a change in temperature causes the gas to expand or contract. These are
usually very accurate in a limited temperature range.

c. Hydraulic thermostat - The hydraulic thermostat consists of a
liquid filled tube (sometimes called a capillary tube) mechanically connected to an electrical contact. When the temperature
changes, the liquid reacts by expanding or contracting, thus
closing or opening the switch contacts.

d. Thermostats are also classified according to usage.

(1) Heating thermostats - Thermostats used to control heating equipment have contacts which close when temperature
drops. When the temperature rises to some predetermined value, the contacts automatically open and disconnect the circuit to the heating equipment.

40

(2) Cooling thermostats - Thermostats used to control fans or other type cooling equipment have contacts that close on temperature rise. When the temperature increases to a preset value the thermostat contacts close and the circuit to the cooling equipment is energized. When the temperature is back down to the desired value, the contacts open and the circuit to the cooling equipment is disconnected.

2. Study the instruction sheet and the Farm-O-Stat Controller and answer the following questions:

   a. What is the difference between a thermometer and a thermostat?

   b. Remove the cover of the thermostat and point out the following parts:

      (1) Nameplate label
      (2) Switch
      (3) Temperature scale plate
      (4) Temperature adjusting knob
      (5) Temperature sensing element

   c. Does electricity connect to anything other than the switch? On this switch, the connection (R) is the common terminal of the SPDT contacts.

   d. What voltages may be applied to this switch?

   e. What is the full load ampere ratings of this switch?

   f. What is the range of temperature adjustment?

   g. What is the differential in degrees F?

3. Perform the following work:

   a. Make the connections by following the diagram on top of next page. Set the temperature dial at 100°G. The motor in this part of the exercise simulates a ventilation fan.

   b. Turn the adjusting knob until the motor starts and runs. When the motor runs, is the room temperature above the temperature setting on the scale plate?

   c. Turn the adjusting knob until the scale plate temperature setting is well above room temperature. Does the motor stop?

4. Perform the following additional work:

   a. Follow the second diagram on the following page and make the circuit connections. The motor will simulate a fan and the light bulb will simulate an auxiliary heating load.

41

Power Supply

120V

Line

SPDT Switch

Fan
Motor

To Load

Thermostat

Power Supply

120V

Line

SPDT Switch

Fan
Motor

Load

Thermostat

Heating Unit

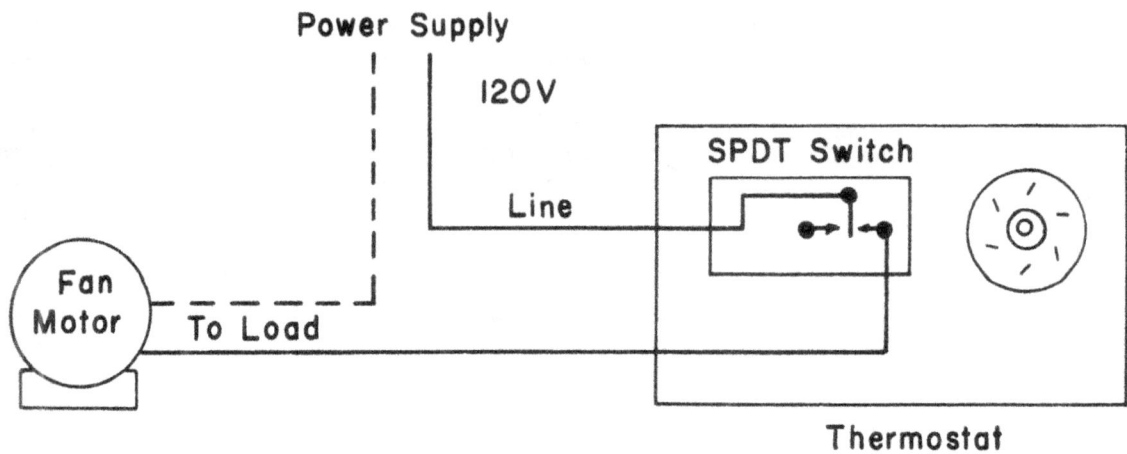

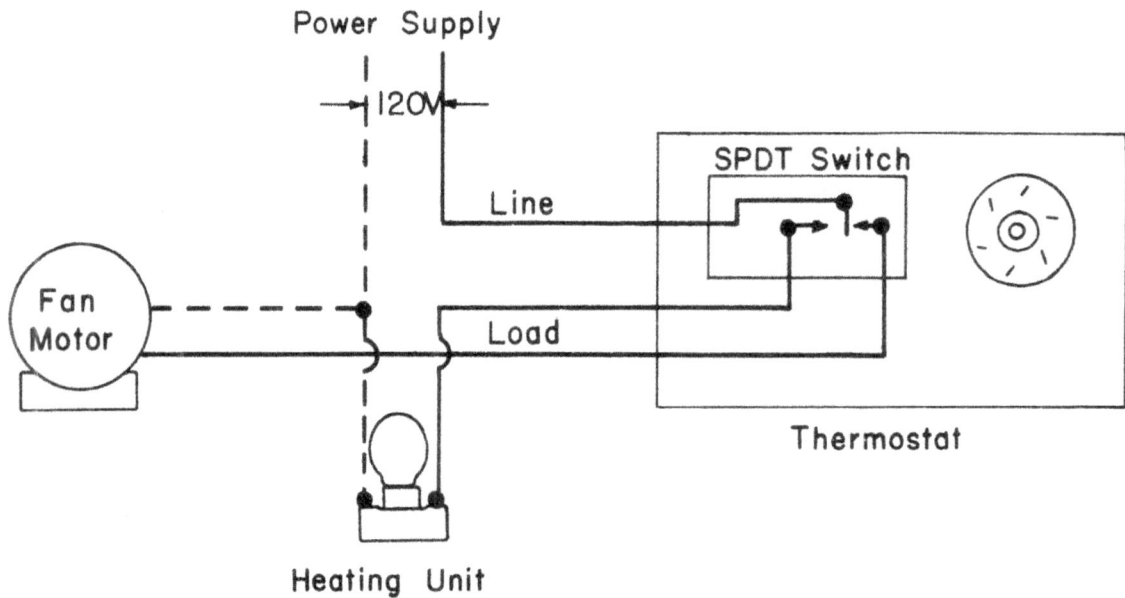

b. Turn the adjusting knob until the scale plate setting is below
room temperature. Does the motor start and run?

c. Turn the scale plate to a temperature setting above room
temperature. Does the motor stop? Does the light bulb come
on?

d. Can the thermostat scale plate be set so that both the motor
and light bulb are off? Why?

e. With the scale plate initially set below room temperature,
carefully turn the adjusting knob until the light bulb comes on.
Don't turn the knob any further than necessary. Now place

the bulb adjacent to the controller sensing element and wait a short while. The bulb should go out and the motor should start and run. Why?

C. Humidity controller - These devices are commonly called humidistats. They utilize human hair or some other material which responds to changes in moisture or humidity. The sensing element is connected mechanically to a set of switch contacts which open or close when the element responds to changes in humidity.

1. Study of humidistat operation - Remove the humidistat cover by pressing in on each side, study the humidity controller, and answer the following questions:

   a. What is the humidity range of this controller?

   b. What is the voltage rating of the switch?

   c. How many full load amperes may be connected directly through the switch?

   d. Point out the sensing element. What is it made of? Some of the humidistats use a human hair for the sensing element. Do not touch any of the internal parts. Explain how humidity affects the sensing element.

   e. Notice the two wires connected to the switch. Is one the neutral? Is the switch installed in the current-carrying (hot) conductor or the neutral?

2. Follow the diagram below and connect the humidity controller and light bulb.

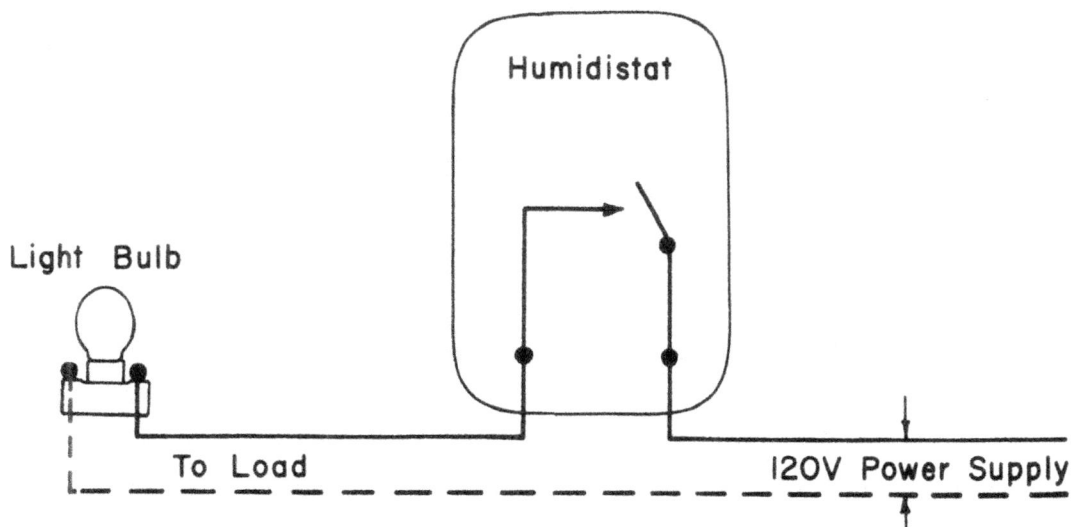

a. Turn the humidistat knob up until the bulb goes out. Don't turn the knob any further than is absolutely necessary.

b. Breathe heavily on the sensing element in the controller. Does the light bulb come on?

c. If you have time, experiment with the humidistat by breathing on the sensing element until the light bulb comes on, then placing the light bulb adjacent to the unit. Be sure that the bulb does not touch the sensing element. As the air temperature rises, the relative humidity should be reduced and the bulb should go out.

3. Perform the following additional work:

a. Connect the humidity controller, the relay, and a motor by following the diagram below.

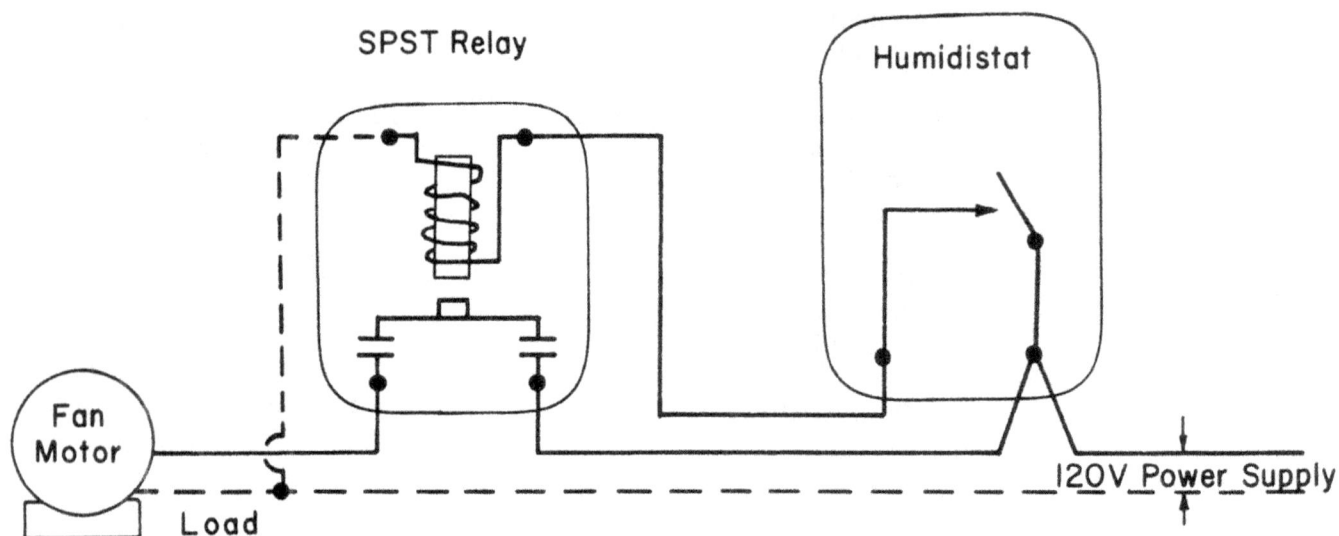

SPST Relay

Humidistat

Fan Motor

Load

120V Power Supply

b. Repeat the same work sequence outlined above.

c. Why is a relay (or magnetic starter) used with the humidity controller when a motor is connected?

d. Can you think of a good farm application for using a humidity controller?

D. Timing devices - These devices are commonly called time clocks or time switches. The timing device usually consists of an electric clock mechanically connected to a set of electrical contacts which are automatically opened or closed at definite time intervals. The frequency of the timing operation may vary from a few seconds to several hours or even days. While there is a wide variety of timing devices on the market, only two will be studied in this lesson plan.

1. General-purpose 24-hour time switch - This device makes one revolution in twenty-four (24) hours and will open and close a switch one time at any pre-set time during the twenty four hour period. Other similar timing devices have removable or adjustable trippers and may be set to open or close a circuit several times during the twenty-four hour period.

2. Repeating cycle percentage timer - This type device is ideal for repeating applications where complete adjustability of the time cycle is desirable. The time interval may be pre-set to close a switch for any part of the complete time cycle. The timer will repeat the "on" or "off" time period as long as power is supplied.

3. Application of timing devices in control circuits

   a. Using the General Purpose 24-hour time switch, perform the following exercises:

      (1) Read the instructions on the inside cover of the time clock.

      (2) What is the "rating" of this time clock? How many amperes may be connected to the load circuit of the switch?

      (3) How do you set the dial for the correct time of day? How many control operations can be performed in a 24-hour period?

      (4) Locate the terminal strip. Refer to the wiring diagram inside the front cover of the case and then point out the following:

         (a) The numbered terminal where the ungrounded (hot) line and load wires would be connected.

         (b) The numbered terminals where the grounded (neutral) line and load wires would be connected.

      (5) The diagram on the following page shows the wiring connections inside the time clock. The wiring connections you will make will be to the terminal strip.

         (a) Connect line wires to the terminal strip. Energize this part of the circuit. Does the clock motor run continuously or would you expect it to run only when the load circuits are on?

         (b) Disconnect the line wires and loosen the trippers on the face of the dial. Set one tripper (on) at 6 p.m. and one tripper (off) at 6 a.m. Set the clock dial at the correct time of day.

Time Clock

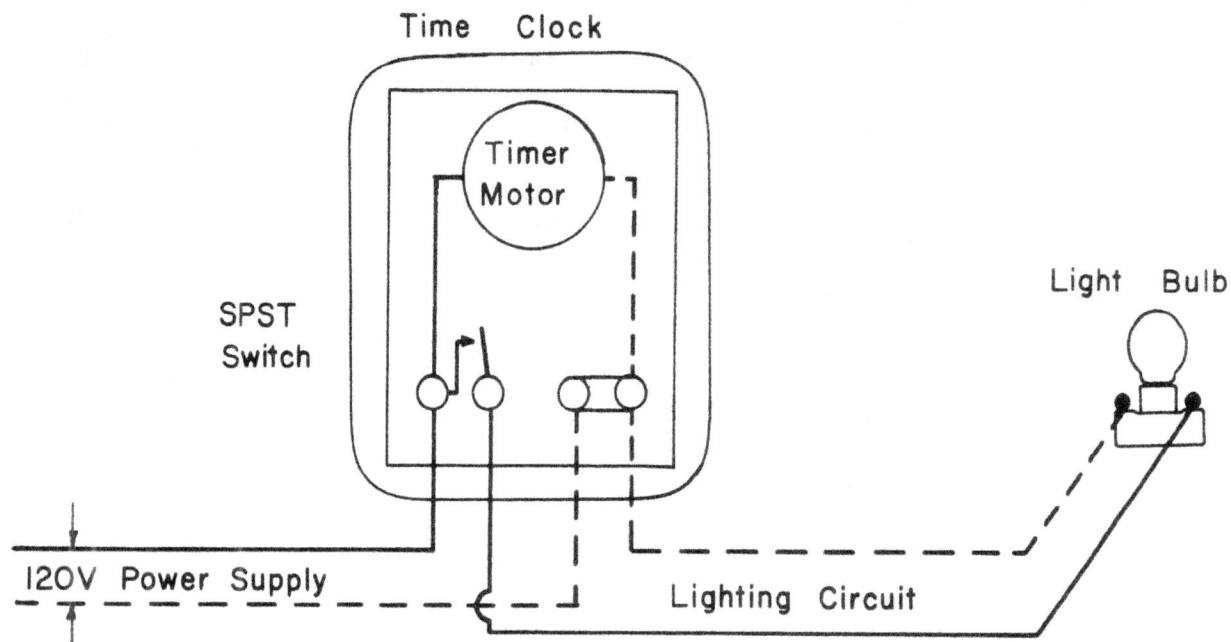

SPST Switch

Timer Motor

Light Bulb

120V Power Supply

Lighting Circuit

(c) Connect a lamp bulb to the load circuit and re-connect the line wires. What happens? Turn the dial until the switch snaps. Does this action turn the lamp on or off? Move the dial until the "off" tripper snaps the switch. What happens?

(d) Problem: We want to have at least 18 hours of light in a poultry house. Set the time clock so that the lights will come on early in the morning and go off at daybreak (6 a.m.) providing 18 hours of lighting (including daylight). Position the trippers and check your settings.

b. Using the Repeating Cycle percentage timer (Paragon Model JW10-0) perform the following exercises:

Some ventilation control systems prefer to have a fan operate for a pre-set number of minutes in a cycle. It may be desired, for example to have the fan operate for two minutes out of every ten-minute cycle. The timer used in this exercise operates on a ten-minute cycle. The number of minutes the fan is to operate during the cycle can be set on the dial.

(1) What is the ampere rating of the switch? What is the voltage rating? Can the switch be connected to 240 volts? What is the maximum size motor that may be connected through this switch?

(2) What is the total cycle time? Each small mark on the dial indicates what time interval? What is the minimum "on" time? What maximum "on" period

may be set on the timer without it being on continuously? Can the fan operate continuously when connected through the timer?

(3) Set the dial to the "off" position. Connect the timer into the motor circuit shown on the following diagram:

Repeat Cycle Timer

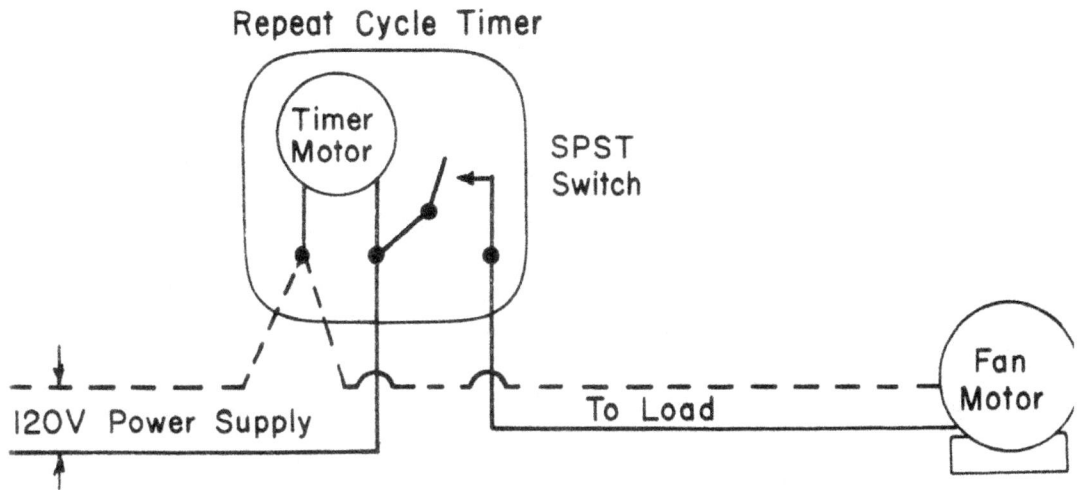

(4) Set a time of two minutes on the timer dial.

(5) Observe the action of the timer for a full ten minute cycle and determine if the fan motor operates for a two minute interval.

(6) Remove the cover from the timer by removing the screw located just below the dial glass and lifting up on the cover. Observe the operation of the working parts. Point out the clock motor. Point out the switch. Is the switch normally open or normally closed? Observe how the lever arm of the switch is caused to depress and release.

(7) Does the clock motor turn continuously? Does it turn when the dial is set to the "OFF" position?

c. It is often desirable to operate more than one automatic sensing device in a control circuit. In the following exercise the repeating cycle percentage timer and a thermostat are both connected into the ventilation control circuit. With this type control circuit the timer can turn the fan on for a short time each cycle to insure the minimum requirements for fresh air even though the thermostat may have the fan shut off because the temperature in the building is below the thermostat setting. On the other hand, if the room temperature rises above the desired setting, the fan would operate regardless of the setting on the timer.

For this type operation the timer and thermostat must be connected in parallel. Follow the diagram below, make the connections, but do not energize the circuit:

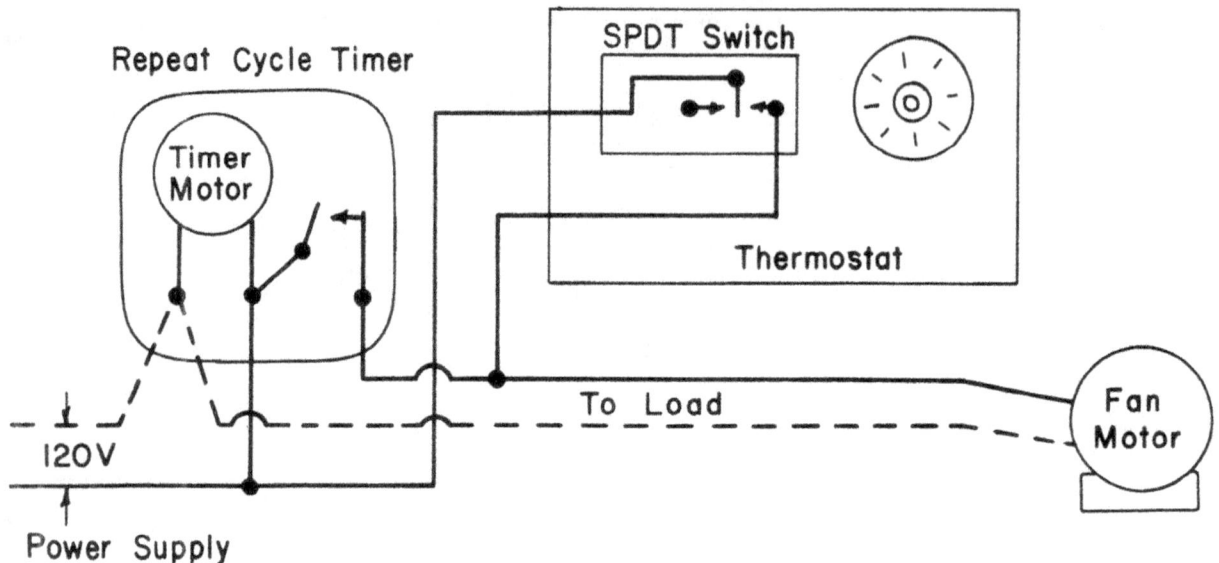

(1) Set the temperature scale of the thermostat to a point above room temperature. Set the timer for a 2-minute interval.

(2) Energize the circuit. The motor should start and run for a 2-minute period after the clock reaches the 10 minute mark. Does it?

(3) After the motor has stopped, move the temperature scale of the thermostat to a point below room temperature. The motor should start and run. Does it? The above action illustrates parallel operation of control elements connected to a ventilation fan.

E. Photoelectric controls - A photocell control is generally used for lighting applications. Street lights are automatically turned on and off by a photocell located in the area. Some yard lightposts have photocell units built in.

By using auxiliary relays or magnetic starters, photocells can also control electric motors. An electrically operated garage door opener, for example, can be activated when a car's headlights strike the photocell. Feed and liquid levels can also permit or stop a beam of light directed at a photocell controlling conveyors or pumps.

Examine the photo relay, study the instruction sheet, and answer the following questions:

1. What is the voltage rating of the unit? How many watts can be connected directly to the unit?

2. Point out the glass lens opening. What do you suppose is the purpose of the louver inside the lens?

3. Is the red colored wire used for connecting the line-in wire or for the load wire?

4. Make the connections as shown in the diagram below:

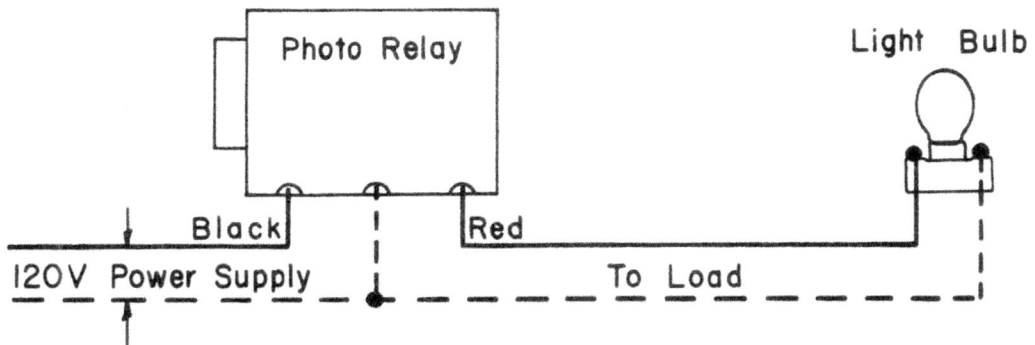

5. Place the photo relay in a well lighted area. Energize the circuit. Cup your hand directly over the glass lens. Does the bulb light or does it go out?

6. What would happen to a controlled light fixture if the lens of the photocell was allowed to become dirty?

7. A photocell of this type is made to operate small resistance type loads (light bulbs) and is not designed to operate induction type loads (electric motors). Therefore, if the photocell is to be used to control a large lighting load (over 500 watts) or operate a motor circuit it will be necessary to use a magnetic relay device in the circuit.

8. Problem - The operator of a beef cattle feed lot desires to provide continuous lighting in the feeding area. The lighting load is 4000 watts and is to come on at dusk and go off at dawn.

    a. What control equipment would be needed?

    b. Draw a schematic wiring diagram showing how the control circuit would operate.

LESSON PLAN - ANSWER GUIDE

Lesson Plan I

Material is self explanatory

Lesson Plan II

A. --

    1. Ampere rating should be stamped on switch.
    2. Voltage rating should be stamped on switch.
    3. Depends upon switch rating as determined by inspection.
       The contact points would burn and damage the switch.
    4. Depends upon the switch being studied.

B. --

    1. Ungrounded wire. If the neutral wire went through the switch, the ungrounded (hot) wire would always be hot to the motor. If the motor then became grounded accidentally, it would run. The Electrical Code requires that switches and fuses be installed on the hot side of the circuit.

    2. --
       a. No. There is no fuse in this switch.
       b. A separate protection device would be necessary. This could be a time delay type fuse or a circuit breaker rated not more than 125% of the full load ampere rating as given on the motor nameplate.

C. --

    1. .87 to .96 amps. Use heater selection chart on switch cover or see manufacturer's catalog.
       a. The heater would get too hot and the switch would trip off to open the circuit.
       b. The switch is rated at 1 HP. The heater unit used, however, is rated at .87 to .96 amps.
       c. Use chart on switch or in manufacturer's catalog. If no charts are available, select a size which is about 125% of the full load motor current rating.

    2. --
    3. --
    4. No. The ampere current flow to the motor goes through the heater. If the heater is removed, the circuit is opened.

5. --
    a. Depends upon size motors connected. Add together the ampere rating of motors.
    b. The circuit should open after a brief time delay.
    c. It should.
    d. Have students time the period.
6. Allow time to cool off and then reset.

D. Starting or stopping augers in feed handling, filling and emptying bins automatically, etc.

1. --
    a. Motor should not run.
    b. Motor should run.
    c. Normally open.

2. --
    a. Motor should run.
    b. Motor should stop.
    c. Normally closed.

3. --
    a. Motor should not run, light bulb should be on.
    b. Motor start and light bulb go off.

E. --

1. Check ampere and motor rating of switches. Switches are not designed for large motor loads.
2. A 3-way switch has three terminals instead of two.
3. --
4. No. Because two circuits may be controlled. One circuit will be "on" and the other circuit "off," depending upon switch location.
5. Both switches.
6. --
7. --
8. Four terminals.
9. --
10. --

Lesson Plan III

A. --

B. --

1. Find rating stamped on relay or check manufacturer's catalog.
2. Check rating on relay or see manufacturer's catalog.
    a. Rating should be indicated on relay coil.
    b. See manufacturer's catalog.
3. Voltage and amperage rating should be stamped on relay.
    a. Multiply volts times amps. Volts x amps = watts.
    b. Check catalog specifications.

        4. --

C. --

D. --

        1. --
        2. Motor should run.
            a. Only one contact is used for 120 volt motor connection. Neutral or ground wire should not go through relay contact.
            b. --
        3. --
            a. --
            b. --
            c. Because current rating of large motors would be greater than could be handled by thermostat switch contacts. Relay controls motor without current going through thermostat.
            d. Controlling ventilating fans in a poultry house.

E. --

        1. --
        2. --
        3. --
            a. Should be apart. Relay is normally open.
            b. --

                (1) Check amperage rating of relay. Volts times amps gives watts. 120 volts x amps = watts.
                (2) Bulb should not come on immediately. Bulb should come on after about 10 seconds.
                (3) Bulb should go off immediately. Because circuit is disconnected and no current is flowing from power source.
                (4) No. A 1/4 HP motor draws more than 3 amps.
            c. --
                (1) --
                (2) --
                (3) Motor should not start immediately. Motor should come on after circuit has been energized for about 10 seconds. Heater in time delay relay closes contacts after 10 seconds. This completes circuit to power relay coil which closes relay load contacts allowing current to flow to motor.

Lesson Plan IV

    A. --

    B. --

    C. --

    D. --

1. --
2. The starter is a three pole switch.
3. Check rating stamped on starter and study manufacturers' catalogs.
4. Starter has overload protectors. One protector is in each hot line of the starter between switch and motor terminals. Select size heater from chart on inside of starter cover.

E. --

    1. --
    2. --
    3. --
        a. --
        b. Contact points close. Start button normally open.
        c. Contact points open. Stop button is normally closed.
        d. 600 volts. Yes. 120/240 volts in most cases.
        e. One terminal of the stop side is connected to one terminal of the start side inside the switch. Only one wire than needs to be brought outside from this common connection. The other two wires brought outside are from the other terminals on the stop and start side.

F. --

    1. --
        a. If wired correctly, all answers should be yes.
        b. If wired correctly, all answers should be yes.
        c. --
    2. --
        a. --
        b. No. Only the small current needed to energize the coil of the magnetic starter.
        c. Since the coil current requirement is quite small (perhaps only a fraction of an amp), small wire can be used for the control circuit even for great distances. The wire must be insulated for the voltage applied, however.
        d. Motor would be protected if the heaters were properly sized. If too much current due to motor being overloaded flows through heater or protective device in the starter, the N/C switch in the holding coil circuit will open, causing the motor starter switch contacts to open, which will disconnect the motor from the power source before it is damaged.

Lesson Plan V

    A. --

    B. --

        1. --
        2. --
            a. A thermometer registers or measures temperature. A thermostat opens or closes a switch with changes in temperature.

b. --

c. No. Electricity connects only to switch of thermostat.

d. 120/240 volts.

e. 7.4 amps on 120 volts and 3.7 amps on 240 volts. Check rating on cover of thermostat.

f. Check dial on thermostat - (35°F to 100°F).

g. Check to determine differential. Should be about 2°F.

3. --

   a. --

   b. Yes.

   c. Yes it should.

4. --

   a. --

   b. Yes.

   c. Yes. Yes.

   d. No. The switch is always closed to circuit regardless of the temperature setting.

C. --

1.

   a. 25% to 70% R.H. (check manufacturer's catalog).

   b. Check voltage rating on humidistat or in manufacturer's catalog.

   c. Check the amperage rating stamped on humidistat or see manufacturer's catalog.

   d. Sensing element is either hair or nylon. The element contracts as the humidity falls and expands as the humidity rises.

   e. No. The switch is installed in the "hot" line only.

2. --

   a. --

   b. Bulb should come on.

3. --

   a. --

   b. --

   c. Motors generally require more amperes then the humidistat switch rating, therefore a relay must be used.

   d. Controlling aeration fans on grain storage bins or foggers in a greenhouse.

D. --

1. --

2. --

3. --

   a. --

      (1) --

      (2) Check instructions and specifications for clock.

      (3) See instructions and specifications for clock.

      (4) --

      (5) --

         (a) Clock motor should run when circuit is energized regardless of load circuit.

b. --
    (1) Check instructions and specifications.
    (2) 10 minutes. One minute. One minute. Nine minutes. Yes.
    (3) --
    (4) --
    (5) --
    (6) Switch is open or closed depending upon the time cycle.
    (7) Clock runs continuously even when dial is set to ''off'' position.
c. --

E. --

1. Check rating stamped on unit and manufacturer's catalog.
2. Louver regulates amount of light required to actuate photoelectric cell.
3. Load wire.
4. --
5. Bulb should come on.
6. Lighting fixture would come on before it should.
7. --
8. Power source, disconnect switch, photo relay, power relay and lighting fixtures.

E E I Publication 66-42

www.ingramcontent.com/pod-product-compliance
Lightning Source LLC
Chambersburg PA
CBHW081749200326
41597CB00024B/4441